Food and Agribusiness Marketing in Europe

Food and Agribusiness Marketing in Europe

Matthew Meulenberg, PhD
Editor

International Business Press
An Imprint of
The Haworth Press, Inc.
New York · London · Norwood (Australia)

Published by

International Business Press, 10 Alice Street, Binghamton, NY 13904-1580, USA.

International Business Press is an imprint of The Haworth Press, Inc., 10 Alice Street, Binghamton, NY 13904-1580, USA.

Food and Agribusiness Marketing in Europe has also been published as *Journal of International Food & Agribusiness Marketing*, Volume 5, Numbers 3/4 1993.

The Haworth Press, Inc., 10 Alice Street, Binghamton, NY 13904-1580 USA

Library of Congress Cataloging-in-Publication Data

Food and agribusiness marketing in Europe / Matthew Meulenberg, editor.
    p. cm.
    Includes bibliographical references and index.
    ISBN 1-56024-474-7 (alk. paper)
    1. Food industry and trade–Europe–Case studies. 2. Farm produce–Europe–Marketing–Case studies. 3. Agricultural industries–Europe–Case studies. I. Meulenberg, Matthew.
HD9015.A2F66 1993
381'.41'094–dc20
                                                                93-34606
                                                                    CIP

INDEXING & ABSTRACTING

Contributions to this publication are selectively indexed or abstracted in print, electronic, online, or CD-ROM version(s) of the reference tools and information services listed below. This list is current as of the copyright date of this publication. See the end of this section for additional notes.

- *ABSCAN, Inc.*, P.O. Box 2384, Monroe, LA 71207-2384

- *Abstracts on Tropical Agriculture (TROPAG)*, Royal Tropical Institute, 63 Mauritskade, 1092 AD Amsterdam, The Netherlands

- *AGRICOLA Database*, National Agricultural Library, 10301 Baltimore Boulevard, Room 002, Beltsville, MD 20705

- *Biosciences Information Service of Biological Abstracts (BIOSIS)*, Biosciences Information Service, 2100 Arch Street, Philadelphia, PA 19103-1399

- *Food Market Abstracts,* Leatherhead Food Research Association, Randalls Road, Leatherhead, Surrey KT 22 7RY, United Kingdom

- *Food Science and Technology Abstracts (FSTA)*, scanned, abstracted and indexed by the International Food Information Service (IFIS) for inclusion in Food Science and Technology Abstracts (FSTA), International Food Information Service, Lane End House, Shinfield, Reading RG2 9BB, England

- *Foods Adlibra*, Foods Adlibra Publications, 9000 Plymouth Avenue North, Minneapolis, MN 55427

- *Produce Marketing Association Database & Bibliography*, Produce Marketing Association, 1500 Casho Mill Road/P.O. Box 6036, Newark, DE 19714-6036

- *Referativnyi Zhurnal (Abstracts Journal of the Institute of Scientific Information of the Republic of Russia)*, The Institute of Scientific Information, Baltijskaja ul., 14, Moscow A-219, Republic of Russia

- *World Agricultural Economics & Rural Sociology Abstracts (CAB Abstracts)*, CAB International, Wallingford Oxon OX10 8DE, England

(continued)

SPECIAL BIBLIOGRAPHIC NOTES

related to indexing, abstracting, and library access services

☐ indexing/abstracting services in this list will also cover material in the "separate" that is co-published simultaneously with Haworth's special thematic journal issue or DocuSerial. Indexing/abstracting usually covers material at the article/chapter level.

☐ monographic co-editions are intended for either non-subscribers or libraries which intend to purchase a second copy for their circulating collections.

☐ monographic co-editions are reported to all jobbers/wholesalers/approval plans. The source journal is listed as the "series" to assist the prevention of duplicate purchasing in the same manner utilized for books-in-series.

☐ to facilitate user/access services all indexing/abstracting services are encouraged to utilize the co-indexing entry note indicated at the bottom of the first page of each article/chapter/contribution.

☐ this is intended to assist a library user of any reference tool (whether print, electronic, online, or CD-ROM) to locate the monographic version if the library has purchased this version but not a subscription to the source journal.

☐ individual articles/chapters in any Haworth publication are also available through the Haworth Document Delivery Services (HDDS).

Food and Agribusiness Marketing in Europe

CONTENTS

∞ ALL INTERNATIONAL BUSINESS PRESS
BOOKS AND JOURNALS ARE PRINTED
ON CERTIFIED ACID-FREE PAPER

ABOUT THE EDITOR

Dr. Matthew Meulenberg is Professor of Marketing and Marketing Research at the Agricultural University Wageningen in the Netherlands and Visiting Professor at the Institute Agronomico Mediterraneo de Zaragoza in Spain. He is also a member of the Board of Directors of the Co-operative Flower Auction Aalsmeer, the Netherlands.

Preface

During the last four decades, a number of changes and developments have taken place in European food and agribusiness marketing. As well, a number of structural and contextual developments have been augmented within the framework of the European Community which created uniform and unified agricultural and food marketing systems with effect from January 1st, 1993. Although a number of legislations have been enacted and agricultural and food management and marketing policies have been developed so far, more work needs to be done in the future for fine tuning of policies and strategies in light of the changes taking place in the volatile European business environment. In particular, problems and hurdles are expected at the implementation stage as there are tremendous divergences among member countries in their needs and expectations as well as their priorities in the implementation of the laws and regulations. Besides these pronounced food and agricultural marketing problems and hurdles among European Community member countries, the intensified trade war between North America and the European Community will have a far reaching impact on food and agricultural marketing in Europe in the years to come. This is creating intra as well as inter trade wars in food and agribusiness areas on both sides of the Atlantic. Problems experienced between Europe and North America may spill over to the Asia-Pacific region thereby creating a struggle among the Triad member countries.

In view of these very important changes and developments taking place in European food and agribusiness, a special volume on the topic was commissioned. Professor Matthew Meulenberg of Wa-

[Haworth co-indexing entry note]: "Preface." Kaynak, Erdener. Co-published simultaneously in the *Journal of International Food & Agribusiness Marketing* (The Haworth Press, Inc.) Vol. 5, No. 3/4, 1993, pp. v-vi; and: *Food and Agribusiness Marketing in Europe* (ed: Matthew Meulenberg) The Haworth Press, Inc., 1993, pp. xi-xii. Multiple copies of this article/chapter may be purchased from The Haworth Document Delivery Center [1-800-3-HAWORTH; 9:00 a.m. - 5:00 p.m. (EST)].

xi

geningen Agricultural University has done an outstanding job as a Editor to develop this special volume for us. In addition to his own Introduction to the volume, there are nine articles on a variety of European agricultural and food marketing issues where some nine European Community countries are represented. In addition to these, an article by Professor Lehota of Hungary presents the characteristics of agricultural marketing in Hungary during the formation of market economy in this advanced developing East European country.

I would like to take this opportunity to thank Professor Matthew Meulenberg for creating such an excellent special volume. I am of the firm opinion that this volume will be an important milestone in the furtherance of a growing body of literature in the area of comparative food and agricultural marketing systems. I offer my heartfelt congratulations to Professor Meulenberg for a job well done.

Erdener Kaynak

Introduction

Matthew Meulenberg

European agriculture, in particular in the EC, has expanded since World War II because of a demand pull and a technology-subsidy push. Immediately after World War II food demand increased because of growing populations and larger incomes. Changes in life style, values and norms influenced the composition of the consumers' food basket, which has shifted from carbohydrates to animal proteins and fresh fruit and vegetables. Improved production efficiency and the CAP have stimulated agricultural production.

But European agricultural markets are changing since the eighties. Food demand is increasing slowly: food consumers are saturated in terms of volume and prefer better to more. Norms and values, in particular with respect to health, environment and animal welfare, are changing. Agribusiness companies and retail chains become bigger, often by internationalization. Large retail chains and alliances of retail companies have developed substantial bargaining power. Innovations in the field of electronic communication, computer technology and biotechnology create new opportunities in production, logistics, information systems and decision support systems. At present, the market situation is also changing because of political changes, like 'Europe 1992,' a changing CAP and a renewal of the GATT treaty.

In connection with these developments in European agriculture

Matthew Meulenberg is Professor of the Department of Marketing and Marketing Research at the Agricultural University Wageningen.

[Haworth co-indexing entry note]: "Introduction."Meulenberg, Matthew. Co-published simultaneously in the *Journal of International Food & Agribusiness Marketing* (The Haworth Press, Inc.) Vol. 5, No. 3/4, 1993, pp. 1-4; and: *Food and Agribusiness Marketing in Europe* (ed: Matthew Meulenberg) The Haworth Press, Inc., 1993, pp. 1-4. Multiple copies of this article/chapter may be purchased from The Haworth Document Delivery Center [1-800-3-HAWORTH; 9:00 a.m. - 5:00 p.m. (EST)].

and agribusiness, agricultural marketing is changing too. A 'State of the Art' survey on agricultural marketing in Europe is timely therefore. We are grateful that many marketing scholars have accepted the invitation to analyze agricultural marketing in their respective countries. This volume is the result of their efforts. It offers a lively picture of European agricultural marketing. It shows both the similarity and the specificity of agricultural marketing in various countries. In this introduction to the volume we will consider in particular the main similarities in European agricultural marketing, as they appear from the contributed papers.

A common characteristic of agricultural markets in various European countries is a stagnating food demand in terms of volume. Population growth is weak. In some countries, like Germany and Belgium, population growth disappeared completely, but in some Mediterranean countries and in Ireland it is substantial yet. The trend towards a lower per capita consumption of carbohydrates and fats and towards a higher consumption of animal proteins seems to reach maturity stage. For instance, in Northern countries like Germany and the Netherlands, consumption of some types of meat is decreasing and that of bread and potatoes is somewhat increasing again. Health considerations and animal welfare arguments play an important role in this respect. The European food consumer appears from various surveys as more quality conscious, also with respect to environmental issues.

Another common characteristic of European agricultural markets is a severe competition between suppliers of agricultural and food products. Competition is severe because of a slowly increasing demand for food combined with a huge agricultural production potential, because of open EC markets, respectively because of internationalization of agribusiness–and of the important role of food retail companies. Food retail companies have substantial bargaining power vis-à-vis food industries because of a large company size and of strong position power in the channel (being the gate to the food consumer). This bargaining power is reinforced by the limited capacity of food producers to be unique in the market by creating unique attractive product properties.

Company size in European agribusiness and food industry has increased in order to realize economies of scale in marketing policies like branding and product innovation. This expansion of multinational food and agribusiness companies is in many countries a challenge to national companies, in particular to farmers' co-operatives, which as a result merge into larger units and often try to build up international positions.

The changes of the CAP, the McSharry plan, and the renewal of the GATT agreement imply, apart from direct income support, more market and less subsidies for EC agriculture. Government support is also decreasing since many governments curtail national support programs, like those in the field of research and extension.

While agricultural markets in many European countries show the same developments, one may also notice many differences: for instance, in particular Dutch and German consumers seem to be much concerned about environmental issues, Spanish agriculture is involved in the transformation towards a modern type of agriculture; Germany is facing the task of integrating East and West German agriculture, Hungarian agriculture is hovering on the brink of a market economy and France along with some other mediterranean countries are in particular concerned about the development of agribusiness.

The major shift in European agricultural marketing, as it appears from various contributions to this volume, is towards more market-customer orientation and, as a result, more concern about product development, branding and customer relationships. Product policy and promotion have become very important in agricultural marketing. While effective and efficient performance of traditional marketing functions—exchange, physical and facilitating—is necessary in every marketing operation, it is not sufficient. Consumer orientation is increasingly the starting point of marketing operations in agriculture and agribusiness. Marketing management has become of strategic importance in European agriculture and agribusiness too. Integrated marketing operations through the food chain are increasingly important in particular for perishables. As a result traditional agricultural marketing institutions, like technical markets, marketing boards, and co-operatives, are searching for maintaining or reinforcing their position by developing marketing strategies which fit to the needs of today's markets. In particular it is important to fit farmers' decisions about the agricultural product properties into customer oriented marketing policies.

Clearly, the stage of development in this trend towards consumer orientation and marketing management differs between products and countries. Some products, like grains, are still of the commodity type and efficient/effective performance of marketing functions–exchange, physical and facilitating–is the core of the marketing operation yet. However in the case of dairy products and fresh horticultural products sophisticated marketing management procedures are needed for an effective marketing operation.

Development in agriculture also differs between European regions, which has its consequences for marketing operations. Nevertheless the contributions to this volume demonstrate that in most countries customer orientation and product, respectively promotional policies, have become important ingredients of European agricultural marketing.

The evolution in the practice of agricultural marketing goes along with the development of agricultural marketing as a discipline. This discipline is shifting from the functional, and institutional approach towards more attention for marketing management. Consumer orientation as a starting point of agricultural marketing operations is stimulating the study of consumer behavior in the agricultural marketing discipline.

Another development in the agricultural marketing discipline is a greater interest in analyzing agricultural marketing as an integrated marketing operation throughout the marketing channel. For that reason vertical marketing systems and other theories and concepts about the structure and functioning of marketing channels become increasingly relevant to agricultural marketing.

Foregoing developments stimulate the development of agricultural marketing towards a multidisciplinary science.

Agricultural Marketing in Germany

M. Besch

SUMMARY. Traditional German analysis and teaching about agricultural markets, "landwirtschaftliche Marktlehre," is concerned with supply, demand and price formation, instruments and policy decisions by policy makers influencing markets and prices, and market structure and distribution.

In this contribution agricultural marketing in Germany will be understood in accordance with the Anglo-Saxon marketing concept. Attention will be paid to: (a) the Marketing environment, in particular food consumption and the competitive position of German agriculture and food industry; (b) the structure of commercialization in particular the agribusiness system and the interrelationships of agriculture with economic sectors supplying to and purchasing from the farmer; and (c) the potentials and limitations of marketing agricultural products. Our paper is primarily confined to the regional states of the former Federal Republic (West Germany). In the description of the marketing environment the present situation in East Germany will be considered too.

INTRODUCTORY REMARKS: SCOPE OF THE PAPER

In the Anglo-Saxon countries 'Food Marketing' comprises 'the performance of all activities involved in the flow of food products and services from the point of initial agricultural production until

---

M. Besch is Professor of Agricultural and Food Marketing, Technical University of Münich, Germany.

[Haworth co-indexing entry note]: "Agricultural Marketing in Germany." Besch, M. Co-published simultaneously in the *Journal of International Food & Agribusiness Marketing* (The Haworth Press, Inc.) Vol. 5, No. 3/4, 1993, pp. 5-35; and: *Food and Agribusiness Marketing in Europe* (ed: Matthew Meulenberg), The Haworth Press, Inc., 1993, pp. 5-35. Multiple copies of this article/chapter may be purchased from The Haworth Document Delivery Center [1-800-3-HAWORTH; 9:00 a.m. - 5:00 p.m. (EST)].

they are in the hands of the consumer' (Kohls, Uhl, 1990). In Germany, however, 'Marketing of agricultural products' is understood in a more specific way, namely the transfer of the Marketing Concept as developed by the industry of consumer goods to agriculture. For the sale of agricultural products, however, the expression 'Vermarktung' is used in science and practice.

The classical 'landwirtschaftliche Marktlehre' as it is taught at German universities, has been developed, in particular by Hanau and his school, from the scientific disciplines agricultural policy and business cycle analysis (Schmitt, 1967). Since its emergence it is a predominantly economic discipline and is focusing on agricultural markets at the macro-level, national or international, whose outcome and structure it is trying to analyze and forecast. Traditionally German 'landwirtschaftliche Marktlehre' is concerned with the analysis of supply, demand and price formation in agricultural markets in order to project future market developments (analyzing and forecasting of the market outcome). The second classical area is the analysis of instruments and of the impact of policy decisions on markets and prices by policy makers. There is a narrow link between both fields of scientific research, which appears from the connection between theoretical market research and empirical agricultural policy in Germany. Actually, the first comprehensive German textbook on 'landwirtschaftliche Marktlehre' was titled 'Agrarmarktpolitik' (Plate, 1968; 1970).

Since the beginning of the sixties a third area of research on 'landwirtschaftliche Marktlehre' has emerged, the analysis of commercialization (market structure analysis and research on distribution). Also in this field problems related to the analysis of the market at the national or international level were dominating, like the analysis of marketing margins, of marketing institutions and distribution channels. In this field of research much attention was paid to vertical integration, contract farming and cooperatives. Finally from this field of research (in the first place by the pioneering activities of Otto Strecker and his colleagues at Bonn University) the agricultural market research has been generated (see Besch, 1981 a, pp. 27).

The reasons for the hesitation of agriculture up till now to adopt the way of thinking and the instruments of marketing, are the special structural and institutional features of the agricultural sector (see Strecker, 1974):

- production by small family farms,
- agricultural products are homogeneous and of the commodity-type,
- the multi-stage organization of agricultural marketing,
- the agricultural policy of government through price- and purchasing guarantees at the most important markets.

These structural and institutional features will have to be taken into account, when trying to introduce the marketing concept into agriculture. In view of the diminishing protection of agricultural regulations as a consequence of the change of the CAP during recent years and because of the increasing competition in the European Common Market recently, the discussion has been intensified under which conceptual and organizational conditions the application of the marketing concept can be improved in the agricultural sector (see Besch, 1990a and coworkers).

In our contribution agricultural marketing in Germany will be understood in accordance with the Anglo-Saxon marketing concept. First of all the marketing environment will be depicted; in particular the changes in food consumption and the competitive position of German agriculture and food industry will be discussed. In the second part of our contribution the structure of the commercialization is described, in particular the agribusiness system and the interrelationships of agriculture with economic sectors supplying inputs to and purchasing outputs from the farmer. The third section of our contribution is dealing with the potentials and limitations of marketing agricultural products.

A treatise on agricultural marketing in Germany after the reunification of Germany in 3-10-1990 requires that also the five new regional states in the former GDR will be considered. Forty years of 'really existing socialism' have brought about a behavior of people and an economic structure, which differ substantially from those in the former Federal Republic. Notwithstanding great efforts and the first advances in the integration of the five new regional states in the market system of the Federal Republic these differences will remain for a long time to come and the harmonization of the economic situation and the conditions of life in both parts of Germany will be realized gradually only.

So at present after being one country for about two years, but having been separated during more than 40 years in two different economic and social orders, a common presentation of agricultural marketing in both parts of the country is not possible yet. Therefore, our paper will have to confine itself in the first place to the regional states of the former Federal Republic (West Germany). In the following description of the marketing environment the present situation in East Germany will be considered too.

THE MARKETING ENVIRONMENT

The marketing environment of the agricultural firms and agribusiness companies, the companies supplying to and purchasing from the agricultural sector, consists in the first place of the changes in food consumption of the population, the competitive position of agriculture and food industry in the Common Market and the influences of market and price policy of the government as related to the markets of agricultural and food products.

Trends in Consumer Behavior

The changes in consumer behavior of the West German population since the end of World War II have been brought about by a strong increase of per capita disposable income and are similar to the well known pattern of quantitative changes of the food demand in western industrial countries (see Wöhlken, 1991a, pp. 44).

In conjunction with a gradually increasing demand for food-energy, carbohydrate rich basic food of vegetable origin (wheat products and potatoes) have been substituted by protein- and fat-rich food of animal origin (meat, eggs, milk products). Parallel to this development, the consumption of sugar, fruit and vegetables has increased. These clear changes in per capita consumption have been satisfactorily explained by econometric methods for the period with a strong economic growth, both for the federal republic and the EC-countries (see Mönning, 1975; Appel/Ferber, 1987).

Already in the second part of the sixties and to a larger extent in the seventies new developments have set out, which cannot be explained by econometric demand analyses, but which have to be attributed to the impact of so called qualitative determinants. For instance, the decrease of butter consumption since 1964, of the egg-consumption since the seventies and the decrease of meat consumption since the beginning of the nineties have been caused by health concerns of consumers (Cholesterol discussion). Since the middle of the seventies the per capita consumption of wheat products in the Federal Republic has been increasing again (see Wöhlken, 1991 b).

Qualitative factors will have an increasing influence on the development of future food consumption in the regional states of the former Federal Republic and with some delay also in the new regional states. Important factors in this respect are pleasure of consumption, the perceived value of the product and the drive for convenience in food preparation and consumption. In addition to considerations with respect to health and safety, the increasing influence of ecological requirements in production, distribution and waste disposal will have an increasing influence on consumers' decisions (Halk, 1992, pp. 63).

Food demand of the *East German population* in the former German Democratic Republic has evolved along lines of economic and social change similar to the West German population since the restoration after World War II: increasing consumption of animal products (meat, eggs, milk products), decreasing consumption of calory-rich products of vegetable origin (wheat, potatoes). Clearly because of the system of economic planning in production, processing, distribution and consumption of food products some characteristic differences showed up in the former German Democratic Republic as compared to the post war evolution in West Germany (see Ulbricht, 1991, Heinz, 1991).

The prior objective of the agricultural and food policy of the German Democratic Republic was food self sufficiency. Imports of food have been of minor importance because of a chronic shortage of foreign currency, while exports have been sporadic only for reasons of earning foreign currency (Heinz, a.o.). Another objective of agricultural and food policy was to serve the population with cheap basic food in order to keep wages and social costs low. By subsidizing food processing (annually about 34 billion GDR-Mark) retail prices were partially fixed below agricultural producer prices. As a result in rural areas substantial quantities of food products have been fed to animals and people behaved carelessly with respect to food products. Therefore figures on per capita consumption in the statistics of the former German Democratic Republic, based on food balance sheets at the wholesale level, overestimate the actual human consumption (Heinz, a.o.). Another characteristic of food consumption in the former German Democratic Republic was the high share of food consumption away from home as a consequence of many outdoor working housewives and the government support for public food supply (Ulbricht, a.o.).

After the monetary union and the reunification, food habits in the eastern part of Germany have become more similar to those in the western part. At present people in East Germany have become more price conscious in food purchasing, waste of food and feeding food products to animals belong to the past and the consumption of food away from home has decreased substantially. Also the composition of the food basket has changed; in comparison to the past East Germans consume presently more margarine and less butter, more fruit and vegetables, more cheese and yogurt and less meat and meat products (Ulbricht, a.o.).

Competitive Position of the Agricultural and Food Business

The competitive position of the German agricultural and food business is determined by two basic developments. At the one side the increasing integration of the German economy in the European inner market, and on the other side the necessary restructuring and reconstruction of the East German agricultural and food business as a consequence of the reunification of Germany. Because of the still existing differences between both parts of Germany and because of the separate data bases, it is necessary to depict the competitive position of German agriculture and food business within the Common Market in two stages: first the competitive position of the West German agricultural and food business will be analyzed vis-à-vis other European competitors. Subsequently, the situation of the East German agriculture and food business within the present stage of reform will be discussed.

The competitive position of West German agriculture in Europe is characterized by small family farms, and as a result a low creation of value per laborer (Zeddis, 1991; Wissenschaftlicher Beirat, 1990). Because of a, at least comparatively speaking, a high productivity per acre, West German agricultural production surpasses in a number of markets domestic needs (wheat, sugar, milk, beef), however, there is a shortage of some products like pork and poultry, oilseeds, fruit, vegetables and wine (see Wöhlken, 1991, pp. 336).

The *impact of the CAP* on this development is evident: surpluses have emerged in particular with products having strong CAP price and intervention regulations. In particular wheat and sugar exports to third countries have been stimulated by CAP export restitutions; surpluses of milk and beef have been exported in the first place to Italy. While German imports of fruit, vegetables and wine largely originate from southern countries, the shortages of animal products based on feed grains (pig meat, poultry meat, eggs) are made up in the first place by imports from the Netherlands (see Wöhlken, 1991 a , pp. 168).

While West German agricultural policy in the context of the EC right from the beginning was focusing on high producer prices and high protection against products from third countries, it is changing in recent years to restoration of market equilibrium by supply restrictions in conjunction with income-allowance for agricultural producers. All in all government agricultural policy has preserved the structure of West German agriculture; the negative consequences of this policy appear not only in the arrears in competitive power of West German agriculture vis-à-vis north western member-countries of the EC, but also vis-à-vis the larger structured East German agriculture (see Prognos, 1991). Presumably in the former German Democratic Republic farms will remain bigger than in West Germany, also after the adaptation of the, out of ideological reasons, strongly enlarged farm holdings in animal- and plant-production to the conditions of a market economy.

The competitive position of the *West German food industry* in the inner EC-market is assessed differently in the field of production and food distribution (see Besch, in Wöhlken, 1991a, pp. 79 and pp. 88). As a result of vertical and horizontal integration, West German food retailing at present is dominated by a small number of large and capable companies, which belong to the biggest international distribution companies. The four greatest German food retail companies are also the four greatest European companies in this field (Prognos, 1991).

However, the German food processing companies are comparatively small according to European standards, let alone as compared to the big internationals. Clearly this picture differs a great deal between various branches of the food industry. While in the dairy industry and in the production of raw materials for bakeries small and medium scale firms are dominating, there is a high degree of concentration in other branches, like the quick freezing industry, sugar industry or the margarine industry, partially because of activities of subsidiaries of multinational companies in Germany (Unilever, Nestlé, Philip Morris). The largest German cooperatives and private meat packing and meat processing industries have acquired a remarkable position in the European market too.

Such different company-size implies a different basis for operation in the future European inner market. The largest West German distribution groups either are represented by subsidiary companies in foreign European Countries (Metro, Tengelmann, Aldi) or belong to international voluntary chains or purchasing groups (Rewe, Edeka, Spar, Markant). In the German food retail business a development can be observed, which already is called 'Eurodistribution,' i.e., concentration of purchasing and logistics at the European level. The opportunities for 'Euro-brands' are considered skeptically by German food producers, also by the internationally operating big companies. The potential advantages in production and marketing (which have also been stressed by the EC-commission as positive aspects of the European inner market, see Besch, 1990) can not be realized as long as food consumption habits of European consumers differ so much. As a result the marketing strategy of the West German food industry will also in the future focus on regional markets, niche marketing, not only because of limited company size but also because of different consumption habits in various European countries (see Weindlmaier, 1991).

The present situation of the *East German agriculture* has emerged from three politically based transformation processes, which have been implemented by force according to the Marxist-

Leninist ideology (Weber, 1991): the landreform of 1946 has lead to expropriation and to the split up of the large estates into smaller farms, the management of a part of the large estates has been continued as state farms 'Volkseigene Güter' (VEG: estates of the people); the collectivisation of 1960 enforced the amalgamation of farms of a village in an agricultural production-cooperative (LPG); the third agricultural reform of the seventies aimed at stimulating agricultural production of the industrial type and led to a radical separation between agronomy/horticulture and animal husbandry. While these two agricultural sectors were brought together in very huge production units, which comprised usually a number (up till 10) of villages. As a result of this policy, roughly 1500 arable farms of, on average, 4500 ha cultivated 92% of the arable land, while 3200 animal farms, practically without land, produced 60-90% of the animal production (see Frenz a.o., 1991, p. 2).

It goes without saying, that the transformation towards the market economy, which followed without delay after the monetary union and the reunification of 3 October 1990, has put a heavy burden on the agriculture of the former GDR. Not only the domestic market (because of the free entry of qualitatively superior products of the west) but also the East European export markets (because of the disappearing currency-free exchange by 'Transfer roubles') collapsed. The absence of selling opportunities has been overcome by massive supportive purchasing by government and by subsidized exports to the former Comecon countries. In the mean time East German agricultural production has been stabilized at a substantially lower level. Decrease in production, either because of agricultural policy (fallowing, milk quota) or by adapting to the new market situation (decrease of feeding potatoes and of human meat consumption), is substantial and varies between 25% (milk) and somewhat beyond 50% (slaughter animals).

In contrast to the wishful thinking of the government, agricultural structure in the former German Democratic Republic after the transition towards a market economy has changed substantially slower than it was hoped for. Up till now about 14000 private farms have been set up, which have an average size of 90 ha (with great differences from north to south), being substantially larger than the average farm size in West Germany of hardly 30 ha (full time farmers). However, since mid 1990 about 3/4 of about 4500 of the former agricultural production co-operatives have been transformed into a new legal structure (co-operative, private company or limited company), together about 3000 firms, which, with an average size of 1400 ha, cultivate about 80% of the arable land of the former German Democratic Republic (Agra Europe 6/91, special issue). As a result only the third stage of the communist agricultural reform has been removed yet, namely the separation of animal and plant production and companies are reduced to the size of one to two villages. Certainly, many of these estates are still in economic trouble and a definitive forecast about the future agricultural structure in the new states is not possible yet.

In strong contrast to the ideologically motivated fundamental transformation of the agricultural structure in the former German Democratic Republic, the field of *processing and distribution of agricultural products* did not change a great deal. Indeed according to the Marxist-Leninist social system, property-rights had been changed and first the industrial companies and the larger trading companies had become public property. Later on a part of craftsmen firms and a part of larger retail companies have been transformed into co-operatives, so that the market share of private companies (inclusive of baking and meat packing) in the distribution of food products amounted to 4% of the value and 10% of the outlets (Frenz a.o., 1991, p. 5).

With few exceptions (i.e., some newly built large plants) food companies were purchasing old fashioned and run down facilities and most of the plants were set up before the second world war. About half of the facilities of food industry in 1987 were worn out according to East German estimates (Heinz, 1991, p. 167). According to West German standards most food processing plants of the former German democratic republic are not competitive any more (only 2 out of the 43 East German sugar factories have been judged fit for renovation by the West German sugar industry; only 5% of the meat packing plants passed the minimum standard for recognition by the EC, see Manegold, 1991).

Public trading companies (HO) and consumer cooperatives participated each with nearly 40% in food retailing, other government companies had a share of 15% while the remaining 4% was in the hands of private retail trade (Manegold, 1991). Food wholesale trade was organized either in co-operatives or in public enterprises. After the reunification the consumer co-operatives remained in existence as independent regional co-operatives, while larger co-operatives have emerged by amalgamation. Their chance of survival is estimated to be small, since their outlets are predominantly small. The other fields of public trade are broken up and privatized to a large extent. By co-operative agreements West German trading companies have partitioned between themselves the former public wholesale trade and large sections of the retail trade of East Germany. In fact only 60% of the East German retail outlets are suitable for modernization, so that a large number of new outlets (it is estimated 40000) will have to be built. Special attention is paid to building mass markets in the outskirts of towns and discount shops in town.

The public East German food industry, brought together in 300 so called 'Kombinate,' has been taken over (like all public companies) by Treuhand. After a difficult period of breaking up and restructuring, Treuhand has been able to sell 404 of the 830 new companies to private buyers, mainly from West Germany. The remaining companies will have to be privatized in the fall of 1992. The increased interest of West German companies in East German firms is brought about according to Treuhand by a renewed interest of East German consumers in food of East German origin (Süddeutsche Zeitung, 20.02.92, p. 31). Clearly this development has been different in various branches. For instance, already quite soon the East German sugar industry has been privatized, almost completely taken over, by the 4 largest West German companies in the industry (Südzucker, Pfeiffer und Langen, Zuckerverbund Nord, Zucker Uelzen/ Braunschweig), which laid their hands in this way on the sugar-quota (see Frenz a.o., 1991, p. 27). The situation is, however, much more complicated in the meat packing and milling industry. These sectors have overcapacity and, in particular in meat packing, completely old fashioned and irreparable facilities (see Agra-Europe 6/92, 03. Febr. 1992, Sonderbeilage).

All in all the privatization of the East German food industry, in particular considering those being taken over by West German companies, will bring about an important investment- and modernization program. Since only a few of the obsolete GDR- companies can be put on a sound basis, there will necessarily have to be built new, large and modern units. At the end of this building phase one has to reckon with the fact that the newly built production facilities in East Germany will be more modern and capable than many of the West German ones. Keeping in mind the possibly cheaper supply of raw materials to these East German processing plants because of the substantially larger agricultural production units in the new regional states, then one should give the East German agricultural and food business good competitive opportunities for the future as compared to the many small scale establishments, in particular in the South, of the former Federal Republic of Germany (see Prognos, 1991, pp. 109).

THE STRUCTURE OF COMMERCIALIZATION

The structure of commercialization in the Federal Republic of Germany should be depicted on the basis of existing links in the Agribusiness. Therefore first a survey of the structure of the agribusiness system will be given, which will be followed by a presentation of the most important branches, participating in the flow diagram of the commodity-relationship. In this context not only the present structure but also the most important developments during the past decades are described. Because of the reasons given (separate developments of both parts of Germany, and missing data bases) our presentation will be confined to the area of the former German Federal Republic (West Germany).

The Agribusiness-System

In an industrial society, based on specialization and division of labor, food provision for the population is realized in a system of interrelated economic branches, which we call after the American 'agribusiness' concept 'food system' (Thimm/Besch, 1971, p. 7). This system can be partitioned into a number of subsystems, the upstream industries supplying products and services to the farmers, the farmers themselves and the downstream industries, processing and marketing agricultural products. The system consists of three structural elements: (1) the branches belonging to the system (institutions), (2) the functions performed in the system and (3) the product flows through the respective branches of the system.

Figure 1 shows the food system in the former German Federal Republic for the year 1990. Because of a limited data base only the turnover of the participating industries are reported; the product flows between the various sectors can be quantified in some cases only (exports and imports, inputs exclusive capital goods and outputs of agriculture).

The first functional phase of the agricultural and food system are the inputs to agriculture. It concerns production means, exclusive capital goods and services for the current production, which amounted to 30 billion Deutsch Mark in 1990. Agriculture also purchases *investment goods* (machinery, buildings) from industries. These expenses for gross investments, which have not been indicated here, amounted to 11 billion Deutsch Mark in 1990.

Sales of agriculture (the turnover of farms) amounted to 54 billion Deutsch Mark in 1990. This figure concerns all products marketed by farmers; they are mainly food products, the share of non food-products (flowers, ornamentals, wool) amounts to nearly 6%.

Food imports of about 60 billion Deutsch Mark–including both agricultural products and processed food products–surpassed the sales of domestic agriculture. Globally the German Federal Republic is one of the most important importers of agricultural products. Certainly also German exports of agricultural and food products, stimulated by the CAP, have increased substantially in the 70's and 80's; in 1990 exports of agricultural and food products amounted to 31 billion Deutsch Mark. German trade in agricultural and food products is primarily intra EC trade: during past years about 70% of German exports and 60% of imports were intra EC-trade.

The lower part of Figure 1 shows the marketing domain of agriculture. This range can approximately be measured by the difference between farm sales and private expenditure for food and beverages, taking also into account the balance of foreign trade in food products. According to their function industries involved in the commercialization of agricultural products can be classified in two

FIGURE 1. Structure of the Food Industry 1990 of the German Federal Republic* (Product flows and sales at different functional stages)

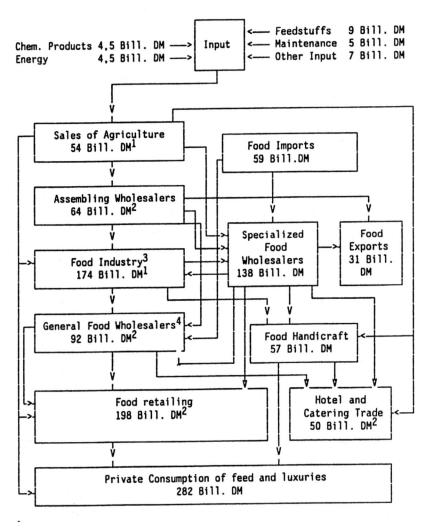

[1] Without Added Value Tax
[2] Taxable turnover
[3] Food industry exclusive feedstuffs; companies having 20 or more employees
[4] Wholesalers of food and luxuries not focusing on specific products
* Former regional countries of the German Federal Republic

groups: companies involved in processing and companies involved in distribution.

The *processing of agricultural products into food products* is the task of the food industry, craftsfirms in food business, and food service business. While craftsfirms essentially consist of bakers and butchers only, food industry can be classified into 28 different groups, which are distinguished either on the basis of the processed raw material or on the basis of the products made. Craftsmen retailers sell self made products in their shops; in addition bakers and butchers increasingly carry other products so that their craftswork as an additional function to the retail function can be imputed to the food distribution apparatus. The Food service (restaurants, etc.) is dominated by the traditional family firm. Nevertheless in this sector–as a result of the disproportion between the increasing demand for and increasing production costs of services–an interesting process of modernization has started, which has been stimulated by the penetration of restaurant chains (Reckert, 1986, pp. 78).

Another interesting point is the two-track system of the *trade in agricultural and food products* in the Federal Republic of Germany. One track consists of the traditional wholesale company, which as assembling wholesaler is specialized on specific agricultural products (agricultural trade, cattle trade), or the food wholesaler carrying only a specific product assortment (e.g., vegetable-, milk-, meat- or beverage wholesaler). The second track of food distribution consists of the food trading company both at the wholesale and retail level which carries a complete food assortment and an assortment of nonfoods. Because of a series of technical and organizational innovations (self service, buying concentration) this track of food distribution has changed profoundly and quickly during the past four decades and its dynamics have had a great impact on the marketing of food and agricultural products.

Structural Changes and Concentration in the Field of Purchasing and Selling by Agriculture

In this section the structural changes and the concentration of the various industries selling to and purchasing from farmers will be reviewed. Similar to the flow diagram of the agribusiness-system our analysis will start out with producers of production means for farmers, afterwards food retail trade and food industry will be discussed as being the most important actors in the agricultural marketing domain.

Industries Selling to Farmers

The value of purchases of production means by West German farmers amounts to more than half of their sales. For that reason it is necessary to pay attention to the markets and marketing of these products. The most important products in this field are compound feed, repair and maintenance of machinery, chemical products and energy (see Figure 1). In most instances agricultural firms do not obtain production means straight from producers but from specialized traders, the so called agricultural middleman ('Landwarenhandel'). Such cooperative and private trading companies do not only sell to farmers but are also involved in marketing agricultural products.

The *compound feed industry* is characterized by both its narrow links with agriculture and with the food industry: their raw materials originate either directly from agriculture or are offals from the food industry. Typical of the West German compound feed industry is its strong dependence on imports, about 3/4 of the processed raw material comes directly or indirectly from foreign countries. However, sales of the West German compound feed industry are mainly domestic. The market structure and competitive situation of the West German compound feed industry has been analyzed by Schwier in 1987. According to this study (Schwier a.o., 1987, pp. 72) in the middle of the eighties 80-90 producers have a share of about 40% of the market and the compound feed factories of 11 central Raiffeisen cooperatives have 30% of the market. Smaller regional wholesale companies and milling companies take care of the remaining 30%.

The *domestic supply of fertilizers* in the Federal Republic is an oligopolistic market, since the production is realized by a few companies having financial relationships (see Pretzel, 1971, pp. 36; Unger, 1984). For instance the market for nitrogen is dominated by two companies: Ruhr-Stickstoff AG (a subsidiary of Norsk Hydro) and BASF. In addition, BASF controls through Kali-Salz-AG practically 100% of the German supply of potassium. The large suppliers of nitrogen, first of all BASF, dominate also the market of compound fertilizer (Henze, 1987, pp. 82). Certainly this domestic monopoly is since the seventies, in particular in the nitrogen market, challenged by foreign suppliers: exports came first from East European countries, and since the middle of the seventies more and more from EC and third countries (Austria, USA).

The *agricultural machinery industry* is composed of two sections, production of tractors and production of agricultural machinery, which often–in particular by the large suppliers–are combined in one company. The market structure for both sections is oligopolistic (see Henze, 1987, pp. 146). The domestic market supply originates from three sources: German producers (Fendt, Deutz-Fahr, Daimler-Benz), international companies, which produce in Germany (Case-International, John Deere, Massey-Ferguson) and domestic suppliers (Fiat). Notwithstanding globally increasing, and also in Germany observable concentration in the industries of agricultural machinery and of agricultural tractors. German companies have been able, in a strongly competitive market, to maintain the position of market leader in various important domestic market segments. With the decrease of the West German investment boom since the end of the sixties, German producers have focused increasingly on foreign markets. Often more than half of the annual production of agricultural machinery and agricultural tractors has been exported during the seventies and eighties. Globally German exports rank second after the USA.

Industries and Trade Purchasing from Farmers

Of all sectors related to the marketing of agricultural and food products the greatest changes have taken place in West German food retailing. Since World War Two this originally very traditional and small scale type of trading has transformed itself into one of the most modern and capable distribution systems of the world through a number of technical and organizational innovations (see Besch, in Wöhlken, 1991a; Besch, 1981 b).

The first of these innovations, which has generated a series of further innovations, has been the *transition towards self service* in food retail trade since the beginning of the fifties. This method of mass retailing has led towards a substantial decrease of labor costs, since some retail tasks have been transferred to the clients (service),

and others to the preceding industry (portioning, packing, information, advice). Like all technical progress, self service implies a
substitution of labor (service personnel) by capital (selling space
and shop furniture).

The further development of self service in West Germany has
expanded into two directions. One direction induced a continuous
increase of store size in conjunction with an expansion of the assortment, which arrived at its tentative completion in the beginning of
the sixties by the emergence of the supermarket, carrying a complete food assortment including fresh products and a smaller non
food section. In the second development direction since the beginning of the seventies the reduction of service, started by self service,
has been further extended by the discount-concept (pioneer: Aldi).
A simple store lay out and product presentation, limited assortment
of convenience products and low labor costs made it possible to
charge prices substantially lower than customary retail prices. The
combination of large retail stores and discounting characterizes the
most recent type of self service in food retailing: the mass market
(Hypermarket). Hypermarkets surpass supermarkets substantially
in store size, the center of the product assortment is already in non
food. In 1988 both modern types of food retailing accounted for
about 40% of total West German food retail sales.

The described changes in the commercial systems (which in
Germany according to Nieschlag are indicated as 'Dynamik der
Vertriebstypen im Handel') have induced a substantial concentration in purchasing in the German food trade. In the fifties and
sixties, concentration of purchasing was speeded up by co-operative
groups in food retailing (purchasing co-operatives Edeka and Rewe
and voluntary chains like Spar and A&O), which could maintain
their position vis-à-vis large food chains. In the seventies and eighties the development in food retailing is determined by new types of
commercialization (discounting and mass markets) and specialized
trading companies (Metro, Aldi, Tengelmann, a.o.). The growth of
large retail chains is not only realized by establishing new outlets at
the cost of the closing down of smaller and non profitable units, but
also by purchasing selling space by take overs of existing companies. In particular because of the penetration of mass markets with a
mixed assortment and because of the financial relationships be-

tween various trade companies, the borders between food and non food retailing have become fluid. The reunification has led to dual growth rates for most West German retail companies. According to estimates of M+M Eurodata, Frankfurt (published in 'Lebensmittel-Zeitung,' 31.01.1992, p. 4) total sales of the German food trading companies amounted to about 300 billion Deutsch Mark (with about 200 billion Deutsch Mark in food and 100 billion Deutsch Mark in nonfood). The ten largest groups of the German food trade commanded about 2/3 of total sales.

The strong vertical concentration of purchasing by large companies in the various stages of wholesale and retail trade has brought about a strong bargaining power of retail companies vis-à-vis their market partners. This bargaining power has its most direct impact on the preceding food industry in the marketing channel.

A general forecast on the competitive position of the *food industry* vis-à-vis the food retail trade is difficult in view of the heterogeneity of this industry. In particular it should be kept in mind that the food processing sector breaks up in industry and craftsmen, which differ in core functions and company size and that companies are classified in different groups on the basis of specific products having a different market position which is determined by a specific supply and demand structure.

On the basis of the market share of the ten largest companies in a sector as the criterion for horizontal integration the 20 most important classes of the West German food industry can be divided into three about equal groups (see Agrarbericht, 1989, Materialband, pp. 106): in 7 classes these ten companies have a market share beyond 80%, in other 7 classes between 50 and 80% and in 6 classes less than 50%. Certainly the groups distinguished in this way are very heterogeneous: for instance in the group characterized by high concentration there are classes with traditional technologies like the margarine- and sugar industry, and industries with relatively new processing methods like the potato processing industry. We find subsidiary companies of international concerns in the first group (margarine industry, oil mills), but also in the middle group (milk processing, fish processing) and in the less concentrated third group (soft drink industry). At the other side there are concentrated sectors in which national firms play a dominant role (sugar industry, potato products, flour industry) (see Wendt, 1988;

Buchholz/Wendt, 1990). Industrial companies doing the first processing of agricultural raw material are also present in all three groups. Clearly businesses of a craft type are more strongly represented in less concentrated business classes.

Economic reality demonstrates, that the degree of horizontal concentration in a sector describes insufficiently the actual competitive situation in markets. This competition is much more–also in the food industry–shaped by the development of sector overstepping mixed companies, which have emerged by interweaving formally independent companies through financial participation (in Germany the international food companies Unilever, Nestlé and Philipp Morris are represented with a number of companies. The largest West German mixed company is the Oetker-Gruppe). The driving forces of this concentration through conglomerates (see also Grosskopf, 1979, p. 25; Breitenacher, 1976, p. 41) are not so much the classic arguments of economies of scale in production and in building market power, but the objectives of profitable investment of liquidities, spreading risks and rationalization of central tasks like R and D. Also logistic advantages and supply-concentration might play an important role in this respect.

Coordination of Buying and Selling in Agribusiness

In contrast to the concentrated markets for buying and selling products in the food business, agricultural production is spread over a large number of small units. As a result there is a poor relationship between the supply and demand structure on the respective stages of the marketing channel, as a consequence of the atomistic demand of farmers vis-à-vis producers of production means and the atomistic supply of agriculture vis-à-vis the concentrated marketing and processing companies. Such divergencies between supply and demand require the performance of functions, which iron out the differences between production and consumption with respect to place, quantity, time and quality.

The connection of agriculture with the preceding and succeeding economic sectors is realized in Germany to a large extent by specialized trading companies, the so called 'Landwarenhandel.' They appear in two legal forms: as private agricultural trading companies (so called 'Privater Landwarenhandel') or as co-operatives (so called Buying and Selling co-operatives). Typically these companies combine purchasing from and selling to the farmer, that means they supply agricultural farms with feedstuffs, fertilizer, pesticides/insecticides sometimes and also with agricultural machinery, and on the other side purchase wheat, oilseed and potatoes. Deliveries to farmers account for 80% of turnover of cooperative agricultural trading companies and for 60-70% of the turnover of private agricultural trading companies ('Landhandelsbetriebe'), purchasing from farmers account for the rest of the turnover (Prusse, 1983, p. 77, 100). Agricultural trading companies are spread over the country in different ways. Private agricultural traders prefer, generally speaking, better locations with large farms, while co-operatives–in view of the responsibility for their members–are more strongly represented in poorer areas having small farms (Prusse, 1983, pp. 124). In conjunction with the decreasing number of farms the number of agricultural trading companies is declining–with the co-operatives–during the past 30 years to 1/5 of the number in 1960.

In selling agricultural products rural co-operatives compete since a long time ago with a great number of specialized private companies in different sectors of industry and trade. The co-operative shares in agricultural sales vary substantially between products and have developed in a different way during the past. According to information of the German Raiffeisen-union the market share of co-operatives is about 70% for milk, 50% for wheat, 33% for slaughter cattle and about 20-30% for fruit and vegetables (DRV. Yearbook- Wendt, 1990, p. 99).

The continuing concentration of purchasing in food retailing has led to a new division of functions in marketing of fresh agricultural products. Product delivery at a buying center requires that market partners make available large uniform quantities of good quality in proper packages. Many small companies, both private and cooperative, in the assembling trade could not cope with this task. As a result the supply of German fresh products might have become less attractive to food retail chains, which shifted therefore to foreign supply.

In some fields of processing agricultural commodities, characterized by a high market risk because of the inelasticity of supply and a high level of fixed costs in production, contract farming as a means of securing raw materials supply has a long tradition (Van Oppen, 1968). By delivery contracts between farmers and processing industry an attempt is made to fine tune between agricultural production and the needs of processing with respect to quality, quantity and time. At the same time middlemen are bypassed in favor of direct contacts between farmer and processing industry. Contract farming in the German Federal Republic in contrast to some other EC countries like Holland and France, up till now has been used to a limited extent. It is in some sectors, like sugarbeets (market regulation!), vegetables for the canning industry and potatoes for processing, poultry and milk, of decisive importance (see v. Alvensleben, 1973, p. 35 and p. 99-Wendt, 1990 b, p. 98).

The disequilibrium between the atomistic supply at the farmers' gate and the increasing concentration in the subsequent sectors requires concentration of agricultural supply too. In the German Federal Republic this concentration has been realized by the establishment of producers' groups according to the market structure law of 1969, which makes it possible not only to improve the production structure by quality regulations, but also a coordination of market supply by suppliers- and marketing groups (Goeman/v.Gruben/Sotzeck, 1969, pp. 285). The number of governmentally recognized producers' groups has reached a peak at the end of 1990 with about 1500, while the number decreased for the first time in 1991. Producers' groups are in particular important for quality wheat, wine, slaughter cattle, piglets and milk (Agrarbericht, 1992, Materialband, p. 134). Though in various regional markets and for some products rather substantial shares in production have been realized (Hülsemeyer/Schmidt/Bunnies, 1977, p. 41, 94, 97), the instrument of producers' groups has not definitely improved the market situation of agriculture.

MARKETING OF AGRICULTURAL PRODUCTS

In the organization of agricultural marketing in the German Federal Republic three different levels can be distinguished:

1. the level of the individual farm,
2. the level of the centralized sector marketing by national and regional institutions,
3. the level of cooperative group marketing by jointly operating agricultural firms.

The marketing instruments can be applied to a limited extent by a farmer, marketing consumption goods directly to the final consumer. In this respect farmers have opportunities in direct contact with their clients to survey the market and to adapt products to the desires of the consumer. Certainly product policy is to a large extent fixed by the natural and economic conditions of farm location. Farmers have a price policy of searching for the outlet offering the best price; farmer's choice of a marketing outlet in direct marketing is limited to two or three alternatives. Also the communication process (advertising and sales promotion) has to restrict itself to low cost activities (see Pottebaum, 1988 and Mahler, 1990). Limited opportunities for marketing by the individual farm leads to the question, whether the performance of specific marketing functions can be transferred to sector institutions.

The first efforts to improve the organizational bases for agricultural marketing in Germany were aiming at removing the structural disadvantages to marketing by farmers and at introducing marketing from the top down by law. The result of these efforts was the decree of the Marketing-Fund Law of 1969, and the establishment of the central marketing company (CMA). In this way it was hoped on the one side to solve the outsider problem and on the other side to develop sufficient competitive power vis-à-vis suppliers of agricultural and food products from other EC countries (see Strecker, 1971). Opposite to the advantage of organizing total agriculture and food industry by law in one central marketing organization there are important disadvantages as well. For instance a central organization should assume a neutral position in competitive matters and has to balance the interests of the various participating economic groups. Also it is natural that a central marketing organization cannot develop product policies, price policies and distribution policies for specific products, because discretionary power in this respect is not with the central organization but with individual companies. The tasks of such central and regional marketing organizations are therefore always supportive and are restricted to marketing service functions like basic- and export market research, basic- and generic advertising, support of sales promotion activities, exhibitions and exchanges, and in the field of product policy labelling and quality control. Important activities are also extension, education and information to the participating industries (Besch, 1981a, p. 31; see also Dallmeier, 1972 Bd. 2).

In view of the limited potential for marketing by individual farms and the restricted marketing domain of the central marketing organization, the middle organizational level of agricultural marketing is very important. Certainly the participation in joint marketing companies is not well developed so far. Therefore further development of agricultural marketing will depend to a large extent on whether one will develop suitable organizational structures for group marketing.

It seems that producers' groups under the Market Structure Law are suited for performing group marketing tasks for their members. The Market Structure Law allows (see Recke/Sotzeck, 1970) producers' groups to develop and monitor regulations with respect to quality and production (i.e., collective product policy), to supply jointly their production (i.e., distribution and acquisition policies); producers' groups have the right of price arrangements with their members (they have their own price policies). Next to these tasks, indicated in the law, producers' groups might also have a task in purchasing production means for the farmers, through subsidiary companies (see Helzer, 1981).

It requires that producers' groups in the first place be quality oriented and being loosely connected, will evolve into tightly organized marketing organizations (see Elsinger, 1991). Such marketing oriented producers' groups might contribute in the framework of co-ordinated marketing systems to collective brands, which differentiate themselves from mass products (Helzer, 1981; Balling, 1990, p. 201). Agricultural production methods and the quality of raw material are becoming more important in view of the increasing health concern of consumers and the increasing demand for organic food. It gives tightly organized producers' groups, which produce according to fixed and controlled production specifications, the opportunity to become the indispensable partner of marketing companies in the context of contractually organized marketing of group brands.

THE CAPACITY OF THE MARKETING SYSTEM

There is no generally valid criterion, which is theoretically based and operational, for the evaluation of the capacity of such a complex system like an agribusiness system in a highly developed industrial country. This holds also with respect to transaction costs theory (Williamson, 1990), which is highly esteemed among West German agricultural economists, but whose empirical test leaves much to be desired yet. One has to resort therefore mostly to the measurement of specific criteria, like marketing margins, speed of innovation or market share. Even for that purpose large scale research projects are necessary as for instance those that have been carried out in the USA in the eighties (see Marion, 1986). In West Germany a huge research project, commissioned by the Ministry for Food, Agriculture and Forestry, on market structure, price development and margins has been carried out by more than 10 research institutions, of which the results have been published by the Society for research in agricultural policy and rural sociology in Bonn (Zurek, 1966). Since that time no comparable research project has been performed in Germany. Only marketing margins for the individual agricultural products are calculated annually by the Institute for Agricultural Market Research of the Research Institute for Agriculture in Braunschweig-Völkenrode and published in the 'Agrarberichten des Bundesministeriums für Ernährung, Landwirtschaft und Forsten.' These figures, calculated as the difference between farm sales and consumer expenditure for food of domestic origin (for the methodology see Wendt, 1986), are reported in Figure 2.

It appears from the data, that the farmer's share in consumer food expenditure is decreasing and marketing margins have increased accordingly. This development holds in general for all food products, albeit its strength differs between products: the decrease of farmers' share has been strong in particular for wheat products, potatoes and meat, but has been smaller, however, for sugarbeets/

FIGURE 2. Share of Agricultural Sales in Consumer Expenditure for Food of Domestic Origin in the Federal Republic Germany in %

	1950/ 1951	1960/ 1961 (a)	1960/ 1961	1970/ 1971 (b)	1970/ 1971	1980/ 1981 (c)	1989/ 1990	1990/ 1991 (d)
Vegetable products (1)	53	41	34	26	33	23	14	12
Wheat for bread and bakery products	46	34	24	13	19	15	8	7
Potatoes	81	70	70	62	63	45	33	30
Sugar Beets and sugar	42	43	39	35	42	42	39	39
Vegetables	37	34	34	27	28	-	-	-
Fruit	67	38	38	38	38	-	-	-
Animal products	68	63	61	55	52	50	43	-
Meat and meat products (2)	69	61	59	48	48	45	36	31
Milk and milk products	66	62	63	65	57	57	53	46
Eggs	81	79	85	85	85	80	71	70
Food Total	64	56	52	48	49	45	36	31

(1) Since 1980/81 without fruit and vegetables.
(2) Meat products inclusive.
(a) Old calculation - (b) New calculation - (c) 1970/71 Most recent calculation - (d) Estimate: results to a limited extent comparable with previous years, because of reunification of Germany.
Comment. In this calculation agricultural sales for food purposes are compared with consumer expenditure for the corresponding food products. For that reason the reported shares are indicative.

sugar, milk and eggs. However, one should not draw premature conclusions on the basis of these different developments of margins about differences in marketing effectiveness. The growth of marketing margins is resulting from two factors (see Wöhlken, 1991 a, p. 123): on the one hand the amount of built in services and on the other hand their service costs. Since the former factor implies an increase in the performance of the marketing system, one can analyze only whether the cost increase as a result of more services is justified by rationalization or not. This question can be answered only by empirical research, which unfortunately is not available for Germany in the present situation.

The assessment of the capacity of the marketing system for agricultural and food products has to restrict itself to some summarizing qualitative statements. It appeared from the description of agricultural marketing that in particular in food retailing, technical and organizational innovations have been carried through rapidly since the Second World War. Also in the food industry new products and new production methods have been introduced rapidly. Even in food service and restaurants unexpected developments have appeared as a result of the introduction of the restaurant chains (e.g., fast food restaurant–see Reckert, 1986). In the craftsman shops (bakers and butchers) only small progress has been made, both in production method and in selling method. This is certainly one of the reasons for strongly expanding margins in the wheat and meat marketing.

Notwithstanding the slow change in farm size, the technical progress of West German agriculture has been very substantial and has caused a large increase of productivity per acre and per laborer. A larger supply increase than the weak increase of demand has brought about a negative trend in the real prices of agricultural products, in particular agricultural products characterized by a strong disequilibrium between supply and demand (see Wöhlken, 1991 a, p. 118).

Since the Second World War increase of German food prices was according to the Central Statistical Office slower than the increase of cost of living. Since in this period–as indicated–marketing margins have widened, it may be concluded that West German agriculture has contributed disproportionally to providing the German population with food at cheap prices.

REFERENCES

Alvensleben, R. v. und Mitarbeiter (1973). Vertikale Integration and Verträge in der Landwirtschaft, Teil 1: Bundesrepublik Deutschland. (Hausmitteilungen über Landwirtschaft, Nr. 106) Brüssel.

Appel, V. und P. Ferber (1987). Vorschätzung des Nahrungsmittelverbrauchs in den Ländern der EG (12) im Zieljahr 1990/95. (Angewandte Wissenschaft, H. 339), Münster-Hiltrup.

Besch, M. (1990a) und Mitarbeiter. Marketing für die Agrarwirtschaft. "Agrarwirtschaft," Jg. 39, H. 9 (Sept. 1990), S. 267-300.

Besch, M. (1981a). Agrarmarketing, Grundlagen und Beispiele. "Marketing," Zeitschrift für Forschung und Praxis, 3. Jg., H. 1 (Februar 1981a), S. 27-36.

Besch, M. (1981b). Strukturwandel auf den Märkten und Bewertung der Kooperation "Ber.üb.LdW," Jg. 59, S. 25-38, Hbg. u. Bln.

Besch, M. (1990b). EG-Binnenmarkt-Auswirkungen auf die bayerische Land- und Ernährungswirtschaft, "Bayer. Ldw.Jb." 67.Jg., SH 2, S. 33-48.

Breitenacher, M. (1976). Untersuchung zur Konzentrationsentwicklung in ausgewählten Branchen und Produktgruppen der Ernährungsindustrie in Deutschland. Hrsg. v.d. Kommission der EG. Brüssel.

Buchholz, H.-E. und Wendt, H. (1990): Ernährungsgewerbe der Bundesrepublik Deutschland im EG-Binnenmarkt. "Landbauforschung Völkenrode," 40, Jg., H. 1, S. 75-87.

Dallmeier, W. (1972). Zentrales Marketing für Nahrungsgüter in der Bundesrepublik Deutschland. (Giessener Schriften zur Agrar-und Ernährungswirtschaft, H. 2 und 3), Frankfurt/M.

Elsinger, M. (1991). Erzeugergemeinschaften als Organisationsmodell zur Förderung eines marktgerechten Agrarangebots. Ergebnisse einer empirischen Untersuchung in Bayern unter besonderer Berücksichtigung des Schlachtviehbereichs. Agrarwiss. Diss., TU München-Weihenstephan.

Frenz, K. u.a. (1991). Zur Agrarmarktsituation in den neuen Bundesländern. IflM-Arbeitsbericht 91/2. Braunschweig, Januar.

Geoman, D., Gruben, H. v. und Sotzeck, M. (1969). Marktstrukturgesetz und Absatzfondsgesetz, zwei neue Initiativen zur Ausrichtung der deutschen Agrarmarktpolitik. Berichte über Landwirtschaft. Bd. 47, 283-300, Hamburg.

Grosskopf, W. (1979). Tendenzen, Ursachen und Wirkungen der Konzentration im Ernährungssektor. In: SEUSTER/WÖHLKEN (Hrsg.), Konzentration und Spezialisierung im Agrarbereich. (Schriften der Gewisola, Bd. 16) Münster-Hiltrup 1979, S. 23-40.

Grosskopf, W. und Alter, R. (1978). Die Marktstellung der landwirtschaftlichen Produzenten. Auswirkungen der Veränderungen bei Ernährungsindustrie, Ernährungshandwerk und Lebensmittelhandel infolge verstärkter Unternehmenskonzentration und internationaler Verflechtung. Abschlussbericht einer Untersuchung im Auftrag des BML. (Vervielfältigtes Manuskript). Göttingen.

Halk, K. (1992). Ursachen des Konsumentenmisstrauens gegenüber Lebensmitteln. Dissertation, Weihenstephan.

Hamm, U. (1991). Landwirtschaftliches Marketing. Grundlagen des Marketing für landwirtschaftliche Unternehmen. (UTB 1620), Stuttgart.

Hausmann, F. (1979). Konzentration und Spezialisierung im ländlichen Genossenschaftswesen der Bundesrepublik Deutschland. In: Seuster und Wöhlken (Hrsg.), Konzentration und Spezialisierung im Agrarbereich. (Schriften der Gewisola, Bd. 16), Münster-Hiltrup, S. 95-113.

Heinz, S. (1991). Stand und Ausblick bei der Verarbeitung landwirtschaftlicher Produkte in der ehemaligen DDR. In: Merl, S. und E. Schinke (Hrsg.), Agrarwirtschaft und Agrarpolitik in der ehemaligen DDR im Umbruch. (Giessener Abhandlungen zur Agrar-und Wirtschaftsforschung der Europäischen Ostens, Bd. 178) Berlin 1991, S. 161-172.

Helzer, M. (1981). Verbundmarketing landwirtschaftlicher Betriebe. Dissertation, Göttingen.

Henze, A. (1987). Die Produktionsmittel der Landwirtschaft. Theorie der Faktornachfrage, Faktoreinsatz und Faktormärkte, Stuttgart.

Hülsemeyer, F., Schmidt, G. und Bunnies, H. (1977). Effizienzprüfung der landwirtschaftlichen Erzeugergemeinschaften am Beispiel des Schlachtvieh-und Qualitätsgetreidesektors. (Landwirtschaft-Angewandte Wissenschaft, H. 200), Münster-Hiltrup.

Kohls, R.L. and J.N. Uhl (1990). Marketing of Agricultural Products, 7th ed., MacMillan Publishing Company, New York.

Mahler, M. (1990). Marketing für Ab-Hof-Verkauf bayerischer Agrarprodukte. Agrarwiss. Dissertation, TU München-Weihenstephan.

Marion, B.W. (1986). The Organization and Performance for the U.S. System. (NC 117 Committee) Lexington u. Toronto.

Manegold, D. (1990). Die Nahrungswirtschaft der DDR im Übergang zu Marktwirtschaft und Binnenmarkt. "Landbauforschung Völkenrode," Jg. 40. H. 4.

Mönning, B. (1975). Nachfrage nach Nahrungsmitteln in der EG (6)-Analyse und Projektion. Agrarwiss. Dissertation, Giessen.

Oppen, M. von (1968). Möglichkeiten und Grenzen der Anwendung vertraglicher Regelungen beim Absatz landwirtschaftlicher Produkte. Hrsg. v. Institut für landwirtschaftliche Marktforschung der FAL, Braunschweig.

Plate, R. (1968). Agrarmarktpolitik. Bd. 1. Grundlagen, München.

Plate, R. (1970). Agrarmarktpolitik. Bd. 2. Die Agrarmärkte Deutschlands und der EWG. München 1970.

Pottebaum, P. (1988). Handbuch Direktvermarktung-Neue Wege für den Absatz landwirtschaftlicher Produkte. Münster-Hiltrup.

Potucek, V. (1988). Strukturelle Wandlungen im deutschen Lebensmitteleinzelhandel und ihre Auswirkungen auf den Wettbewerb, Berlin.

Prettzell, M. (1971). Möglichkeiten der Marktbeobachtung, Marktanalyse und Marktprognose auf landwirtschaftlichen Beschaffungsmärkten, dargestellt am Düngemittelmarkt der Bundesrepublik Deutschland. Diss.agr., Bonn.

Prognos (1991). Analyse der Position der bayerischen Land- und Ernährungswirtschaft im Marktbereich für Agrarerzeugnisse und Erarbeitung effektiver, zukunftsorientierer Marketingstrategien. Untersuchungsbericht, Basel, Mai 1991.

Prüsse, D. (1983). Der Landwarenhandel in der Bundesrepublik Deutschland. Struktur der Wettbewerbssituation im privaten und genossenschaftlichen Landwarenhandel (Schriftenreihe des BMELF, Reihe A. Angewandte Wissenschaft, H. 291) Münster-Hiltrup.

Recke, H.-J. und Sotzeck, K.M. (1970). Marktstrukturgesetz mit Erläuterungen und Materialien. Hildesheim.

Reckert, G. (1986). Zur Adoption neuer Speisen und Verzehrsformen. - Die Einführung von fast food in der Bundesrepublik Deutschland -. Dissertation (oec.troph.), TU München - Weihenstephan.

Schmitt, G. (1967). Zur frühen Geschichte der landwirtschaftlichen Marktforschung in Deutschland. In: Schmitt, G. (Hrsg.) Landwirtschaftliche Marktforschung in Deutschland. München-Basel-Wien.

Schürmann, K. (1984). Landwirte und deren Handelspartner - Ein empirischer Beitrag zur Marktstellung. (Forschungsgesellschaft für Agrarpolitik und Agrarsoziologie Nr. 269). Bonn 1984.

Schwier, D. (1988). Struktur und Wettbewerb auf dem Markt für Mischfutter in der Bundesrepublik Deutschland. "Agrarwirtschaft," SH 114, Frankfurt/M. 1987.

Strecker, O. (1971). Gemeinschaftsmarketing für Nahrungsmittel. "Agrarwirtschaft," Jg. 20, H. 9, S. 281-285.

Strecker, O. (1974). Agrarmarketing. In: Tietz, B. (Hrsg.) Handwörterbuch der Absatzwirtschaft, Stuttgart.

Strecker, O., J. Reichert und P. Pottebaum (1990). Marketing für Lebensmittel. Grundlagen und praktische Entscheidungshilfen. 2. überarbeitete Aufl., Frankfurt/M.

Thimm, H.-U. und Besch, M. (1971). Die Nahrungswirtschaft. Zunehmende Verflechtung der Landwirtschaft mit vor- und nachgelagerten Bereichen. (Agrarpolitik und Marktwesen, H. 12) Hamburg und Berlin.

Ulbricht, G. (1991). Ernährung und agrare Veredlungswirtschaft in der früheren DDR. "Agrarwirtschaft," Jg. 40, H. 5, S. 134-138.

Unger, H. (1984). Nachfrageanalyse des EG-Düngemittelmarktes. (Agrarmarktstudien, H. 29) Hamburg und Berlin.

Weber, A. (1991). Zur Agrarpolitik in der ehemaligen SBZ/DDR. Rückblick und Ausblick. In: Merl, S. und E. Schinke (Hrsg.), Agrarwirtschaft und Agrarpolitik in der ehemaligen DDR im Umbruch. Berlin, S. 53-70.

Weindlmaier, H. (1991). Konsequenzen des EG-Binnenmarktes für den der Landwirtschaft nachgelagerten Bereich. In: Schmitz, P. M. und H. Weindlmaier (Hrgs.), Land- und Ernährungswirtschaft im Europäischen Binnenmarkt und in der internationalen Arbeitsteilung. (Schriften der Gewisola, Bd. 27), Münster-Hiltrup, S. 481-488.

Wendt, H. (1990a). Der Lebensmittelhandel in Ländern der EG: "Landbauforschung Völkenrode," 40. Jg., H. 1, S. 88-92.

Wendt, H. (1990b). Massnahmen und Instrumente zur Marktstrukturverbesserung und Absatzförderung in der Bundesrepublik Deutschland vor dem Hintergrund des Gemeinsamen Binnenmarktes. "Landbauforschung Völkenrode," 40. Jg., H. 1, S. 93-105.

Wendt, H. (1986). Anteil der Verkaufserlöse der Landwirtschaft an den Verbrauchersausgaben für wichtige Nahrungsmittel inländischer Herkunft in der Bundesrepublik Deutschland - Berechnungskonzept und Ergebnisse. "Landbauforschung Völkenrode," 36. Jg., H. 2, S. 79-88.

Williamson, O.E., (1990). Die ökonomischen Institutionen des Kapitalismus. Unternehmen, Märkte, Kooperationen, Tübingen.

Wissenschaftlicher Beirat beim Bundesministerium für Ernährung, Landwirtschaft und Forsten (BMELF) (1990). Agrarpolitische Konsequenzen der Realisierung des EG-Binnenmarktes bis 1992. (Schriftenreihe des BMELF, Reihe A. Angewandte Wissenschaft, H. 384) Münster-Hiltrup.

Wöhlken, E. (1991b). Strukturwandel im Lebensmittelverbrauch. "Agrarwirtschaft," Jg. 40, H. 5, S. 133/134.

Wöhlken, E. (1991a). Einführung in die landwirtschaftliche Marktlehre. (UTB 793) 3. Aufl. Stuttgart.

Zeddies, J. (1991). Wettbewerbsfähigkeit aus der Sicht landwirtschaftlicher Betriebe - Situationsbeschreibung und Auswirkungen in ausgewählten Harmonisierungsbereichen. In: Schmitz, P. M. und H. Weindlmaier (Hrsg.), Land- und Ernährungswirtschaft im Europäischen Binnenmarkt und in der internationalen Arbeitsteilung. (Schriften der Gewisola, Bd. 27) Münster-Hiltrup, S. 75-87.

Zurek, E. (1966). Marktstruktur, Preisentwicklung und Spannen bei ausgewählten land- und ernährungswirtschaften Erzeugnissen in der Bundesrepublik Deutschland. Zusammenfassender Ergebnisbericht. (Forschungsgesellschaft für Agrarpolitik und Agrarsoziologie, Nr. 160) Bonn.

Strategic Marketing Objectives in Mergers and Acquisitions in the Greek Food Industry

George G. Panigyrakis

SUMMARY. The purpose of this study is to provide empirical evidence of the effect of the type of ownership (foreign vs. domestic deals) on the objectives of corporate acquisitions in the Greek food industry, by reviewing the current literature and then analyzing a selected group of recent takeovers. In order to determine whether or not any effect did in fact exist, a set of twenty acquiring objectives were used on a number of possible strategic, marketing, financial and managerial goals. Analysis of these data indicated that, contrary to what would be expected from previous research Greek food mergers differ according to the nature of the deal (domestic vs. foreign).

INTRODUCTION

As a result of the development of the Single European Market, there has been an important merger and acquisition (M&A) activity in all EEC member States. Despite the fact that this activity reflects more overall trends in mergers and acquisitions in all members,

George G. Panigyrakis, MSc and PhD in Food Marketing, is Associate Professor of Marketing at the Athens University of Economics and Business.

The author acknowledges helpful comments on an earlier draft from Professor M. Meulenberg.

[Haworth co-indexing entry note]: "Strategic Marketing Objectives in Mergers and Acquisitions in the Greek Food Industry."Panigyrakis, George G. Co-published simultaneously in the *Journal of International Food & Agribusiness Marketing* (The Haworth Press, Inc.) Vol. 5, No. 3/4, 1993, pp. 37-54; and: *Food and Agribusiness Marketing in Europe* (ed: Matthew Meulenberg), The Haworth Press, Inc., 1993, pp. 37-54. Multiple copies of this article/chapter may be purchased from The Haworth Document Delivery Center [1-800-3-HAWORTH; 9:00 a.m. - 5:00 p.m. (EST)].

37

there has been an underlying increase in cross-border activity. The same trend is present in Greece and in the Greek food industry in particular.

The topic of this article focuses on the M&A activities in the Greek food industry investigating whether they aim at the general industrial restructuring in the food sector of Greece, in response to the challenge of the Single European Market, their objectives, motives and success. In this paper we argue that Greek food mergers will be different from those of the major EEC States. For, the benefit of integration will be different because the motivation is more likely to be the prospect of entering into new markets, rather than financial factors or the exploitation of economies of scale or experience curves. The costs of integration will be different because the pattern of ownership and control of companies is different as well as the acquisitor's access to information on prospective target food companies.

These findings have important strategic, tactical and internal marketing implications. Strategic implications affecting the form of integration (joint venture, minority share stakes or full merger). Tactical implications influencing the way that one has to choose a partner and internal marketing implications with respect to the particular human resource issues that may arise in the effort to mobilize the participation of the two members to pursue a common marketing strategy.

PARTICULARITIES OF THE MERGER MOTIVES/BENEFITS

The wave of mergers and reserve mergers in recent years has attracted important attention, but most of the academic and public discussion has been devoted to the mergers' consequences. The motives behind these mergers have received only modest attention although they ultimately decide whether a merger is tried or not.

M&A motives have initiated far less theoretical research than merger consequences. Research in this area has taken two general approaches.

The first approach is to indicate a comprehensive list of the various managerial motives and objectives that might motivate managers to engage in M&A activities (Steiner, 1975; Allen et al., 1981; Goldberg, 1983; Howell, 1970).

The second approach is to focus on how particular management goals or objectives motivate managers to engage in M&A activities. This approach views mergers and acquisitions as being planned and practiced to:

1. Achieve synergies in the form of financial, operational and managerial synergies (Porter, 1985; Jensen & Murphy, 1988; Rumelt, 1986; Montgomery & Singh, 1984; Scherer et al., 1975; Kitching, 1967; Jensen, 1986; Friedman & Gibson 1988; Maremont & Mitchell, 1988; Chatterjee, 1986; Eckbo, 1986; Shelton, 1988).
2. To achieve market power (Porter, 1985; Steiner 1975; Ellert, 1976; Chatterjee, 1986; Scherer, 1980; Rhoades, 1983; Jensen, 1984; Feinberg, 1985; Ravenscraft & Scherer, 1987).
3. To use privileged market information (Steiner, 1975; Holderness & Sheehan, 1985; Ravenscraft & Scherer, 1987, Wensley, 1982).
4. To maximize the top management's own utility instead of the shareholders' value (Baumol, 1959; Mueller, 1969; Rhoades, 1983; Black, 1989; Porter, 1987; Dunkin, 1988; Friedman & Gibson, 1988; Rothman, 1988; Smith & Sandler, 1988; Bartlett, 1988; Dobrzynski, 1988; You & Al., 1986; Walsh, 1988).

THE GREEK EXPERIENCE OF M&A IN THE FOOD INDUSTRY

The absence of a relevant European code has forced European firms in their international M&A activities to respect their specific national antitrust regulations. Most commonly M&A activities have assumed the form of a firm acquiring another by buying its stock (Katsos & Lekakis, 1990). This arrangement ensures an easy entry and exit, saving different divestiture costs when the partnership proves unsuccessful. Another type of partnership which is considered as a partial merger is the joint venture, that is a quite common practice in the EEC. The major features of M&A activity in the EEC are summarized in Jacquemin et al., 1989. Domestic mergers dominate over international ones and are prominent among large companies.

Greek Antitrust Policy, which is expressed in just a few legislative acts, has restricted companies from conducting any other strategic activity but merger, and acquisition.[1]

Legislation has been used traditionally by the Greek policy makers as a means of strengthening economic progress through the developments of large companies capable of exploiting economies of scale (Katsos & Lekakis 1990). Thus, instead of restrictions to M&A activities one finds different incentives consisting of tax allowances granted to the post-merger company,[2] as mergers and acquisitions are not restricted by an important antitrust legislation. Despite this favorable climate, M&A transactions in Greece did not occur until the late 1980's and the early 1990's (Figures 1 & 2, Tables 1 & 2).

METHODOLOGY

A list of 20 possible objectives was developed for decision makers engaged in M&A activates. This list was derived from previous research by Kitching, 1967; Howell, 1970; Steiner, 1975; Walter and Barney, 1990 and is shown in Table 5. Particular care was taken to include possible managerial goals from several different disciplines, including marketing, economics, finance, organization theory and strategy.

Responses were collected from 54 executives of the acquired food firms, having in mind that they may not be completely free from certain biases as have been suggested by Walter and Barley, 1990. As for meaningful comparisons of domestic vs. foreign deals in food M&A objectives the technique of matching was used to control independent variables which might affect the results. Thus, in order to eliminate the effect of a possible influential independent variable on the dependent variable, the subjects were chosen so that they were as homogeneous as possible on that independent variable. As a result the subjects were matched on an organizational level and

1. Law 703/1977.
2. L.D. 4002/1959 and L.D. 1297/1972.

FIGURE 1. M & A deals in Greece (1983-April 1992)

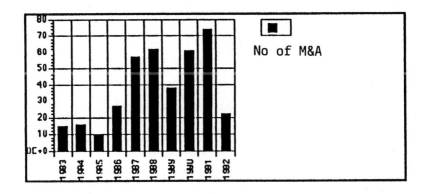

*1992: Jan-April.   Source: ICAP

FIGURE 2. M & A transactions by type of deal (foreign vs. domestic)

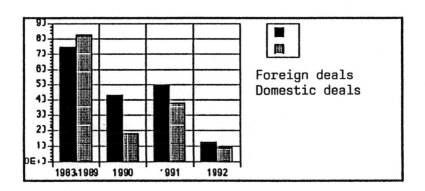

1992: Jan.-April.
Foreign deals: deals between a foreign and a Greek company.
Domestic deals: deals between two Greek companies.

TABLE 1. Major foreign M&A deals in the Greek food sector

```
----------------------------------------------------
Jacobs/Suchard          Swiss              2
Nestlé                  Swiss              1
Group Benedetti         International      1
BSN                     France             2
Pernod/ Ricad           France             1
Pepsico                 U.S.A.             2
3E (Coca-Cola)          International      3
Frand Metropolitan      Great Britain      2
----------------------------------------------------
```

Source: ICAP 1992.

TABLE 2. Nationality of the foreign deals in the Greek food sector   (1987-1991)

Country	Number
France	15
Great Britain	11
Italy	10
Holland	10
Germany	7
U.S.A.	7
Swiss	5
International	7
Other	10

Source: TOP Invest 1992.

type of job. They were all holders of important senior executive positions in their firms (e.g., President, Vice-President, Chairman, Marketing Manager). This procedure provided an objective criterion for selecting subjects and preventing the confounding of the variables of time on the job, education, number of previous employments, etc. Chi-square analysis indicated that there were no differences on critical variables of the subjects (age, sex, education and income). Data were collected through structured interviews by each participant, and were approximately 1-2 hours in duration. The interview focused on ranking the importance of the 20 different objectives for M&A listed in Figure 2 for two different categories of M&A (domestic versus foreign deals).

Following the lead of similar studies, each participant was asked to rank managers' most important objectives for each of the two types of mergers. For each ranking task, participants were given a randomized set of 20 cards with the M&A objectives listed in Table 3. Participants then sorted the cards, selecting and ranking the five most important objectives for everyone of the two M&A types in question. To familiarize themselves with the 20 goals listed in Table 3, as well the ranking task, participants were first asked to rank the importance of these goals of M&A in general. These data were not included in the present study. After this introductory ranking task, each participant ranked the importance of different objectives for each of the two types of M&A (domestic/foreign deals).

The data collected in this process consisted of two 54 by 20 matrices. Each cell in the matrix contained a participant's ranking of the importance of a particular management objective (scale 1-5, 1 = first most important . . . 5 = fifth most important, 0 = not among the five most important) for a particular type of M&A.

FINDINGS

Cluster analyses were performed on the collected data to develop a taxonomy of M&A objectives (Hartigan, 1975). Several different cluster analyses were performed, all generating a high level of consistency across the different clustering results, suggesting quite solid results (Breiger et al., 1975). Spearman's rank correlation coefficient was used to compute the rank order of preferences between domestic and foreign deals. A chi-square test and a t-test were used for most of the cases for independent measures, as well an F-test to assess the homogeneity of variance assumption between domestic and foreign deals.

This study reports in detail only the analysis where all ranks were recorded in a binary form (1 = a motive was ranked among the top five most important by a respondent, 0 = otherwise) and rankings for M&A in general were not included. Ranking of all two types of M&A were considered simultaneously in developing the taxonomy of objectives. The clustering algorithm used was CONCOR (Walter

TABLE 3. List of management objectives for M&A in the Greek Food Industry

---

1. Accelerate growth or reduce costs and risks.
2. Improve competitiveness in obtaining a sizeable market share.
3. Improve the penetration base for existing products of the acquiring company.
4. Create economies of scale by improving the capacity capability.
5. Divest poor performing elements of the otherwise undervalued acquired company.
6. Expand capacity at less cost than assembling new facilities, or physical assets.
7. Fulfill particular objectives of the acquiring company's chief executive.
8. Gain important or potentially valuable assets with the cash flow or other financial strengths of the acquiring firm.
9. Gain complementary financial features.
10. Improve efficiencies and reduce risk in the supply of specific goods to the acquiring firm.
11. Improve economies of scale by using the acquiring form's distribution capacities to absorb expanded output.
12. Penetrate new markets by using the acquired firm's marketing capacities.
13. Promote visibility with investors, bankers with the aim to other benefits in the future.
14. Pursue opportunities to sell stock at a profit.
15. Reduce risks and costs of entering a new industry.
16. Reduce risk and costs of diversifying products and services delivered to customers within an industry.
17. Using synergistic qualities of the acquired company vis-à-vis the acquiring company.
18. Using financial strengths of the acquired company.
19. Utilize the acquiring company's experience and expertise in marketing, production, or other areas within the acquired company.
20. Utilize the acquired company's personnel, skills, or technology in other operations of the acquiring company.

& Barney, 1990) although other algorithms generated quite similar results. The ALPHA cutoff was 0.95. CONCOR results for these ranking data are presented in Table 4.

The application of various statistical tests suggested that four clusters adequately grouped the data in the present analysis. To examine how the importance of these four different clusters of objectives varied by type of M&A, the cluster solution was applied to each of the data matrices for domestic and foreign type of ownership (Greek/foreign). This was done by rearranging the rows of each of these data matrices to match the partition of objectives obtained in the cluster analysis. The average ranking of all participants in each cluster for each type of merger and acquisition shows the percentage of respondents who thought that the goals in a given cluster of objectives were among the five most important for a given type of merger or acquisition. The resulting averages are presented in Table 5 and the rank importance of objectives in Table 6.

MANAGEMENT OBJECTIVES IN M&A IN THE GREEK FOOD INDUSTRY

Table 6 made explicit the pattern of management objectives. The objectives grouped together in cluster One suggest that M&A are a mechanism for managers to obtain and exploit economies of scale and scope. In each case, these objectives focus on taking skills and/or assets in one firm and using them in the other firm to create economies of scale. This cluster of objective is similar to the efficiency oriented arguments of Echbo (1983), Halpern (1973) and others.

The objectives grouped in cluster Two suggest that M&A are sought by managers who are mainly motivated to expand their current product lines and markets. None of the objectives involved in this cluster emphasizes product market expansion. This product market expansion objective underlying M&A has not received much importance in previous research even though it is related to both market power arguments and efficiency arguments.

The objectives in cluster Three suggest that M&A are a way in which managers maximize and utilize a firm's financial capabilities.

TABLE 4. Clustering results

Cluster number	Objective numbers	Description of cluster (objectives)
I	19, 4, 20	Mergers are a way managers obtain and exploit economies of scale and scope.
II	2, 16, 12, 11, 3, 6	Mergers are a way managers expand current product lines and markets.
III	8, 15, 7	Mergers are a way managers maximize and utilize financial capability.
IV	13, 18, 9, 14	Mergers are a way managers deal with critical and ongoing interdependencies with others in a firm's environment.
V	1, 17, 10	Mergers are a way managers enter new business.

TABLE 5. Percentage of times objectives of specific clusters are cited as important for different types of M&A

Cluster number

Merger type	I	II	III	IV	V
Domestic ownership	0.611 'H'	0.579 'H'	0.310 'M'	0.299 'M'	0.072 'L'
Foreign ownership	0.385 'M'	0.081 'L'	0.293 'M'	0.522 'H'	0.388 'M'

47

TABLE 6. Rank importance of objectives in Domestic and Foreign deals

Objectives	Domestic	Foreign	X2
I. Mergers are a way managers obtain and exploit economies of scale and scope.	25	16	0.883
II. Mergers are a way managers expand current product lines and markets.	24	4	11.571
III. Mergers are a way managers maximize and utilize financial capability.	18	14	0.086
IV. Mergers are a way managers deal with critical and ongoing interdependencies with others in a firm's environment.	16	18	0.596
V. Mergers are a way managers enter new business.	5	24	16.467

Everyone of these objectives focuses mainly on the exploitation of capital assets to gain economic advantage. This objective is consistent with the work of Howell (1970), Chatterjee (1986) and Lowellen (1971).

Cluster Four suggests that M&A are a key mechanism through which managers deal with important and ongoing interdependencies with firms in their environment. This cluster is similar to the motives for M&A emphasized by transaction cost and resource-dependence scholars (Williamson, 1975; Pfeffer & Salancik, 1978).

The objectives grouped together in cluster Five suggest that M&A provide a way for managers to enter new business. These results are consistent with the work by Rumelt (1974) and Pitts (1977).

OBJECTIVES IN M&A TYPES OF OWNERSHIP (DOMESTIC/FOREIGN)

The relative importance of each objective cluster for the two different M&A types is shown in Table 5. The overall average of the importance of objectives for the two different types of M&A was 0.31 with a standard deviation of 0.21. Following the lead of Breiger et al. (1975), a designation of 'H' (for high) was given to any cell average that is one standard deviation or greater above the overall mean. A designation of 'L' (for low) was given to any cell average one standard deviation or more below the overall mean. Averages between one standard deviation below or above the mean were designated by an 'M' (for medium).

The findings from this analysis were quite interesting. For 'Greek-domestic' M&A, objectives forming clusters One (mergers is a way managers obtain and exploit economies of scale and scope) and Two (mergers are a way managers maximize and utilize financial capability) are dominant. Cluster Five is considered of less importance (mergers are a way managers enter new business). The other clusters of objectives are moderate. For 'Foreign' M&A, cluster Five emerges as dominant, cluster One, Three, and Four as moderate and cluster Two as unimportant.

In studying the aggregate responses of domestic and foreign deals, a correlation coefficient of –0.133 was found (Table 6). A chi-square test $\times 2 = 28.602$ was calculated and was proven significant at $p = 0.01$. When the rows where examined three of the five contrasts were significant at the traditional acceptable levels (t-value), clusters One, Two and Five. The Spearman's $r(s)$ measure of rank correlation was computed to measure the degree of association between domestic and foreign deals in the rank order of importance in which they place their objectives. A value $r(s) = -0.589$ was calculated, indicating a high degree of dissimilarity between the two different types of deals (no monotonic decreasing relationship in the rank order of the importance they attach to the M&A objectives).

DISCUSSION

In the run up to 1992 there has been an important increase in merger activity in the Greek food industry. Our findings suggest that foreign M&A deals are different from domestic deals.

The objectives of integration in foreign deals are different because the motivation is usually the prospect of entry into new markets with all the synergies that correspond, rather than financial factors or the exploitation of economies of scale. As far as the Greek food industry is concerned, if substantial scale economies are a goal, as the present study suggests, foreign firms have to realize that these usually require heavy post-acquisition investments. Furthermore, no matter the origin of the deal the success factors in Greek food industry acquisitions remain the following: good purchase price, reasonable sales synergies and clear potential for economies of scale.

These observations have important implications for strategy: partial integration, through joint venture or minority share stakes is preferable in the case of foreign M&A to full mergers. Second, the actual choice of a partner is different as the objective is both to identify partners who are capable of facilitating entry into new geographic markets and who will also have the incentive to do so.

In contrast the most important stated factor for domestic mergers is general 'expansion.' However, expansion alone is not a good reason for merger, for while it presents a motivation for combining firms it does nothing to suggest that actual performance will be improved. Its records, at doing so, are not encouraging, as countless studies of previous rounds of merger activity have shown (Meeks, 1977). Expansion remains a good objective only if one believes that it builds or reinforces competitive advantage. However, as most of the domestic deals in the Greek food sector are horizontal ones, executives feel that such acquisitions, when well planned, frequently represent low risk, cost-effective routes to increase market share or improve profitability in the pressure of the domestic and foreign competition.

Market entry that seems the predominant objective in foreign mergers is not, in its turn, a source of competitive advantage. However, it can contribute in revealing the value of the relative competitive advantages that either of the firms may have, by enabling the advantages to be extended to new markets. In this respect, mergers motivated by the desire to enter new geographic markets appear to be directed more toward the exploitation of potential synergies than their domestic counterparts, and it is not surprising that multinational diversity has a stronger link to profitability than product diversity (Grant et al., 1988; Davis et al., 1991).

Finally, while this paper duplicates and extends previous research on particular managerial objectives in M&A, it is also consistent with the 'list' approach. The study shows that often the pattern of multiple objectives is important in M&A.

CONCLUSIONS

This paper adds further insights into the objectives behind M&A in the Greek food industry. The study suggests that domestic M&A activities in the food sector in Greece are used as a defensive response intended to compensate for unattractive prior circumstances. Faced with low market shares and in highly concentrated industries, the sample firms acquired new business in industries where they previously produced output (horizontal integrations). The dominant hypothesis tested in this study indicated that domestic and foreign deals are quite different. Future research might focus productively on the following:

• Are mergers in the Greek food industry more (or less) successful than in the EEC states?

- Are mergers of certain size firms more (or less) successful?
- Are hostile takeovers more (or less) advantageous to the share holders than non hostile ones?
- What is the link between expected performance and long term actual performance of acquisitions?

Regardless of the direction that this future research may take, there is an obvious need to improve our knowledge about why food and drink companies engage in M&A behavior in today's conditions of the Single European Market.

REFERENCES

Allen, M. G., Oliver, R. A. & Schwallie, E. (1981). "The key to successful acquisitions," Journal of Business Strategy, 2, pp. 14-24.

Bartlett, S. (1988). Is RJR work $25 billion? New York Times, 2 December p. A1.

Baumol, W. (1959). Business Behaviour, Value and Growth, MacMillan, NY.

Black, B. (1989). Bidder overpayment in takeovers, Stanford Law Review, Spring, pp. 21-37.

Breiger, R., Boorman, S. & Arabie, P. (1975). "An algorithm for clustering relational data, with applications to social network analysis and comparison with multidimensional scaling," Journal of Mathematical Psychology, 12, pp. 328-383.

Chatterjee, S. (1986). "Types of synergy and economic value: the impact of acquisitions on merging and rival firms," Strategic Management Journal, 7, pp. 119-139.

Davis, E., Shore, G. & Thomson, D. (1991). Continental mergers are different. Business Strategy Review, Spring, pp. 49-70.

Dobrzynki, J. (1988). Are RJR and P. Morris putting stockholders first? Business Week, 7, November, p. 34.

Dobrzynki, J. (1988). "Was RJR's Ross Johnson too greedy for his own good?" Business Week, 21, November, p. 95.

Dunkin, A. (1988). "Hamish Maxwell's big hunger." Business Week, 31, October, pp. 24-26.

Echbo, E. (1986). "Mergers and the market control: the Canadian evidence," California Journal of Economics, 19, pp. 236-267.

Ellert, J. "Merger, antitrust law end stockholder returns," Journal of Finance, pp. 715-732.

Feinberg, R. (1985). "Sales at risk: A test of the mutual forbearance theory of conglomerate behavior." Journal of Business, 58, pp. 225-241.

Friedman, A. & Gibson, R. (1988). "P. Morris Co. is bidding $90 a share for craft Inc. in $11 billion tender offer." Wall Street Journal, 18, October, p. A3.

Goldberg, W. (1983). Mergers: Motives and methods. Nichols, NY.

Holderness, C. & Sheeham, D. (1985). "Raiders or saviors? The evidence on six controversial investors." Journal of Financial Economics, 14, pp. 555-579.

Howell, R. (1970). "Plan and integrate your merger," Harvard Business Review, 49.

Jacquemin, A. & Ilzkovits, F. (1989). Horizontal Mergers and Competition Policy. European Economy No 40, Commission of the European Communities, Directorate General for Economic and Financial Affairs, pp. 33-43.

Jensen, M. & Murphy, K. (1988). "Performance pay and top management incentives." Harvard Business School Working Paper.

Jensen, M. (1984). "Takeovers: Folklore and science." Harvard Business Review, 62(6), pp. 109-121.

Katsos, G. & Lekakis, J. (1990). "Trends and causes of mergers and acquisitions in Greece," Spoudai, vol. 41, No. 1, pp. 26-39.

Kitching, J. (1967). "Why do mergers miscarry? Harvard Business Review," 45, pp. 84-101.

Lewellen, W. "A pure financial rational conglomerate merger," Journal of Finance, pp. 521-545.

Maremont, M. & Mitchell, R. (1988). "Pillsbury could be a grand coup for Grand Met." Business Week, 17, October, p. 30.

Meeks, G. (1977). Disappointing marriage: a study of the gains from merger. University of Cabridge, Department of Applied Economics, Occasional paper, 51, Cambridge University Press.

Montgomery, C. & Singh, H. (1984). "Diversification strategy and systematic risk." Strategic Management Journal, 5, pp. 181-191.

Mueller, D. (1969). "A theory of conglomerate mergers." Quarterly Journal of Economics, 83, pp. 643-659.

Pfeffer, J. & Salancik, G. (1978). The External Organizations: A Resource Dependence, Harper, NY.

Pitts, R. "Strategies and structures in action," Academy of Management Journal, pp. 197-208.

Porter, M. (1985). Competitive Advantage. Free Press, NY.

Ravenscraft, D. & Scherer, F. (1987). Mergers, Sell-Offs and Economic Efficiency. The Brooking Institute.

Rhoades, S. (1983). Power, Empire Building and Mergers. Lexington, MA.

Rothman, A. (1988). "Maxwell hungers to widen P. Morris's brands." Wall Street Journal, 19, p. B10.

Rumelt, R. (1986). Strategy, Structure and Economic Performance. Harvard Business School Press, Boston.

Scherer, F. (1980). Industrial Market Structure and Economic Performance. 2nd edn., Houghton Mifflin Boston.

Scherer, F., Beckenstein, A., Kaufer, E. & Murphy, D. eds, (1975). The Economics of Multi-Plant Operation. Harvard University Press, Cambridge, MA.

Shelton, L. (1988). "Strategic business fits and corporate acquisition: empirical evidence." Strategic Management Journal 9, pp. 279-287.

Smith, R. & Sandler, L. "P. Morris track record in acquisitions draws flak." Wall Street Journal, 20, October, p. C1.

Steiner, P. (1975). Mergers: Motives, Effects, Policies. University of Michigan, Ann Arbor, MI.

You, V., Caves, R. Smith, M., and Henry, J. "Mergers and bidders" wealth: Managerial and strategic factors." In Thomas L. (ed.). the economics of Strategic Planning. Lexington Books, pp. 201-221.

Walsh, J. (1988). "Top management turnover following acquisitions." Strategic Management Journal, 9, pp. 173-183.

Walter, G. & Barney, J. (1990). Research notes and communications management objectives in mergers and acquisitions, Strategic Management Journal, 11, pp. 79-86.

Wensley, R. (1982). "PIMS and BCG: New horizons or false dawn?" Strategic Management Journal, 9, pp. 173-183.

Williamson, O. (1975). Markets and Hierarchies: Analysis and Antitrust Implications. Free Press, NY.

The Agrifood System in Italy: Structural Adjustments to Face the Internationalization of the Food Industry

Franco Rosa
Giovanni Galizzi

SUMMARY. In this article the authors describe the most important changes of the Agrifood system in Italy during the last twenty years. The paper describes the structural transformation of Agriculture and the effort of food industry to adapt to the internationalization of food markets. Agriculture and food industry are conceived of as two components linked by functional relations of the Agrifood system. The structural changes of Agriculture are expressed by a consistent decrease in occupation and contribution to the GNP.

However, structural changes vary greatly between regions and can be explained by a different degree of specialization, concentration of resources, technologies and rural development. Because of structural inefficiencies (small scale, low concentration and integration) and poor commercial strategies, food industry has encountered some difficulties. However, progress has been made during the eighties to cope with these changes.

Franco Rosa is Associate Professor, Institute of Agricultural and Food Economics Facoltà di Agraria, Universita Cattolica del Sacro Cuore, Piacenza. Giovanni Galizzi is Full Professor, Chair of Agricultural and Food Economics, Director of the Institute, Facoltà di Agraria, Universita Cattolica del Sacro Cuore, Piacenza.

[Haworth co-indexing entry note]: "The Agrifood System in Italy: Structural Adjustments to Face the Internationalization of the Food Industry." Rosa, Franco, and Giovanni Galizzi. Co-published simultaneously in the *Journal of International Food & Agribusiness Marketing* (The Haworth Press, Inc.) Vol. 5, No. 3/4, 1993, pp. 55-81; and: *Food and Agribusiness Marketing in Europe* (ed: Matthew Meulenberg), The Haworth Press, Inc., 1993, pp. 55-81. Multiple copies of this article/chapter may be purchased from The Haworth Document Delivery Center [1-800-3-HAWORTH; 9:00 a.m. - 5:00 p.m. (EST)].

55

INTRODUCTION

Considerable attention has been devoted in recent years to the structural and institutional transformation of the agrifood sector in Italy. This sector represented since the early stage of industrial development a convenient source of food and provided a significant economic contribution to the domestic economy (Rosa a.o., 1991).

During the sixties, the industrialization of Italy caused consistent socioeconomic changes. South to North and Rural to Urban migrations accelerated the transition from rural to an urban society. In the period 1963-70 the rural population declined by 1.5 million units from 5.3 to 3.8 million and in the following twenty years another 1.5 million left rural areas (Fanfani a.o., 1991; Galizzi, 1981; Giacomini a.o., 1991).

These are the premises of the agrifood transformations: the decline in economic and social importance of the agriculture, the growing demand for food with correlated imports, of which milk and meat are the most important, a domestic food industry that in spite of its expansion is not able to cope with the rapid expansion of demand. Consequently the food deficit grew during the period 1965-75, and tended gradually to slow down in more recent years. However, in 1987 it was superior to the value of oil imports, in 1988 it amounted to 18,000 billion lire, 60% of which was due to the import of agricultural products.

Political programs to face the growing demand of food products and to stimulate the structural adjustment of agricultural and food sectors were represented by "Piani Verdi" a complex of concerted actions promoted by the State administration to support the agricultural sector, and the constitution of mixed private-public groups with intervention of SME (Societa' Meridionale di Elettricita') and SOPAL (Societa' Partecipazione Alimentare) to promote the development of the food industry.

The recent events represented by the concentration of enterprises that operate in the food industry, the growing internationalization of the food markets, the completion of the common European market, foster an analysis of the economic and political factors that will stimulate the competitiveness of domestic companies in the international food market (Bertelè, 1989; Fabiani, 1984; Galizzi, 1981; Swinbank a.o., 1983).

THE CHANGES IN THE AGRIFOOD SYSTEM

The transition from agriculture to the agrifood system can be described by the following changes:

i.   The change of the traditional farming system, like the exit from agriculture of a substantial number of farmers, a slight increase of the average farm size and a growing integration of agriculture with the food industry in the most advanced regions.

This pattern is very different from region to region and is correlated with the degree of evolution of food and manufacturing industry and more in general with the economic level of the region. In some works this has been defined as a dualistic development which is strongly affected by the imperfections of labor and capital markets (Fabiani, 1984; Fanfani a.o., 1991; Magni, 1982).

ii. Modernization of the food industry by innovation in food technologies, integration and concentration of enterprises to gain competitive advantages (Cannata a.o., 1988; Frigero, 1978; Loseby a.o., 1992). Foods are more and more becoming the synergetic combination of agro-industrial processes and marketing strategies; the typical functions of the food industry: processing, packaging, conservation, transport added to the commercial and marketing functions performed by retail companies are giving to food products the quality standards, accessibility and image attributes, required by modern distribution. Dairy products, processed fruit and vegetables, meat and poultry, grain milling and chocolate, are characterized by rapid demand expansion due to product innovation and changing retailing techniques.

iii. The expanding organization of the Agrifood system.

This is evident in the most developed regions of Po River Valley: Piemonte, Lombardia, Veneto, Emilia-Romagna, with the emergence of "Agrifood districts" characterized by vertical coordination among enterprises operating in agriculture, food and manufacturing industry, wholesale and retail distribution. Concentration and specialization are particularly advanced in the meat, vegetable and wine sectors where a network of manufacturing enterprises supplies the technologies and differentiated services to develop competitive products (Balestrieri, 1988; Bagarani a.o., 1988; Iacoponi, 1990; Ievoli, 1986; Terrasi, 1985).

iv. The concentration of consumer demand and its evolution, as a consequence of the urbanization, has created new market opportunities at the retail level and has increased the importance of commercial strategies.

The growing number of professions in the tertiary sector, women's emancipation, smaller size of families, decline in traditional values and the importance of communication to influence individual choices are the most important factors determining remarkable changes in food demand in Italy (Bertelè, 1980; Chang Ting Fa, 1987).

Emerging tendencies in food consumption are represented by:

1. An increasing demand for convenience foods;
2. Growing quantity and expenditure for food consumed away from home express lifestyle changes as a consequence of new occupations and higher per-capita income.
3. The imitation of consumption habits of other countries which implies the globalization of consumer demand (Barkema a.o., 1991).

The increased elasticity of consumer demand and the profit realized at the retail level demonstrate that the marketing strategies are able to change the traditional relationship between food and consumer. Consumer motivations are increasingly determined by a combination of physiological, psychological and economic factors, that represent the interaction between the consumer and his environment: education, profession and media are becoming very important determinants of food choice. In this context, product price and income are two but not all the determinants of consumer choice. The globalization of consumers' tastes, that have determined the market expansion of products like Citrus juices, Olive oil, Pasta, Pizza, Parmesan style cheese, is responsible for the introduction of foods that are not in the Italian style like margarine as a butter substitute, seed oil for olive-oil, cereal derived products, fro-

zen foods, French cheese and wine, vegetables from other countries, tropical and out of season fruits.
4. The negative externalities caused by indiscriminate use of pesticides, fertilizers and chemicals for the intensification of agricultural processes.

The alteration of the original quality of natural resources, sometimes associated with a mediocre quality of mass production, has determined the growing opposition of different groups against the negative consequences for the environment and consumer health of intensive use of pesticides, chemical and some biological inputs: hormones, antibiotics and other products used in intensive farming.

The number of works published in Italy in the last ten years demonstrates the academic and professional interest in and the acquisition of a deeper knowledge about this subject. Theoretical and institutional problems, political implications and marketing strategies have been extensively examined in the literature, in order to explain the role and changes of the Agrifood system.

Galizzi (Galizzi, 1987) has developed a vertical integration approach framed within an Industrial Organization scheme to study the consequences of structure, conduct and strategies of the enterprises operating in the Agrifood system. This approach emphasized the role of the food industry in determining the functional relations that link the food industry to farming and distributive sectors. He pointed out the structural weakness of the Italian food industry that reflects the structure of manufacturing industry, the small size of the farms, and the specific attitudes of Italian consumers to allocate a consistent share of their food expenditure in traditional foods regionally or locally produced in small, family managed enterprises. The structure of the farming sector was in many cases responsible for technical and financial inefficiencies of the food chain: in 1988, with an average dimension of seven hectares, significantly lower than the European average, the 2.14 millions of units below five hectares represented 77% of the total and only 22.7% of the arable land. Structural inefficiency, low farmer education, ageing and rural isolation represent the difficulties for the agricultural sector to make a consistent structural adjustment to the strategies of the food industry. The size limits of the food enterprises may explain the acquisition and control of some important Italian food groups by multinational ones. Many of these are interesting business, which despite their small size as compared to the size of international companies are performing quite well due to the 'niche position' (niche for product, customer or channel) achieved in the domestic market. This 'niche position' makes them able to cope with the greatly segmented Italian demand for food products. Sometimes linked to domestic stereotypes nevertheless the niches or specialty food markets are the most recent evolution of food marketing. This was possible with the progress in food technologies enabling farmers and food processors to target consumer niches more precisely and to adapt the commercial strategies of the distribution according to specific market conditions. The demand for a wider variety of foods demonstrates the greater interest for specific food attributes: nutrition, health, image, convenience.

THE ECONOMIC IMPORTANCE
OF THE AGRIFOOD SECTOR

The INEA (INEA, 1991) estimated for the 1990s the global value of the Agrifood sector at 190 billion dollars of which:

i.   19%, the value added of agriculture,
ii.   8%, the intermediate consumption of agriculture,
iii.   10%, the amount of agrifood investments,
iv. 14.5%, the value added of the food industry,
v. 48.5%, residual value of commercial and distribution services.

The growing importance of the food processing industry is signalled by its increased value added compared with the primary sector.

In Table 1 are reported for agriculture, food and manufacturing sectors (i) the value added at factor costs (VAFC) expressed in current national values (ii) the share of VAFC with respect to the total GDP.

The economic importance of the three sectors generally declined, indicating the transition from a prevailing secondary to a growing tertiary economy. In this period agriculture showed the greatest

TABLE 1. The contribution (VAFC) of agriculture, food and manufacturing sectors to the national GDP: Absolute values, Period 70-89. (Bil)

Year	Agriculture Abs.val	%	Food processing Abs val.	%	Manufacturing Abs.val.	%	Total GDP Abs. val
70	5356	8.73	1766	2.88	16587	27.04	61348
75	10121	7.68	3615	2.74	36111	27.41	131738
80	22631	6.20	10270	2.81	98948	27.10	365173
85	39586	5.20	17835	2.34	193610	25.45	760769
89	46805	4.27	25140	2.29	272882	24.88	1096815

Source: Our elaboration of data from National Institute of Statistics (ISTAT)

decline, loosing more than half of its contribution to the GDP, the food sector lost 20%, more than 9% decrease of the manufacturing sector. The decline of agriculture was systematic during the period while food processing and manufacturing experienced in particular a decrease of their relative contribution to GDP in the period 1980-85. Despite this apparent decline, the importance of agriculture with 4.3% of GDP, is remarkably higher than the 3.2% of the EEC, 2.1% of the United States and 3% of Japan.

Three signals of the declining importance of agriculture are:

i.   The demand for set-aside: 680,000 hectares representing 53% of the total EEC set-aside programme.
     In Emilia Romagna, a productive agrifood region, the set-aside program caused in the period 1988-1992 a growth of the land withdrawn from production from 930 (at the beginning) to 15,000 hectares (at the end) and the number of farms involved in the programme has grown from 94 to 1983.
ii. The reaction of producers to the EEC programs of supply control, in particular the milk quota control. In most of the Padania regions responsible for 85% of the domestic milk production, the number of farms that will stop production in the next years will grow.
iii. The declining Italian competitive power in Citrus, Olive oil, Tomato markets, due to a better marketing organization of exporting companies in Spain and Israel and cost advantages, specifically labor costs, of North-African countries and Turkey.

The decline of agriculture is expected to accelerate in the coming years due to lower revenues of a large number of small farms, which face growing costs especially labor costs, being less protected by CAP price policies, due to the growing bureaucracy required for the new compensation policies and because of the emerging difficulties of integration with food industry.

Unfortunately policy makers prefer to continue in myopic conservatism to maintain their influence, instead of facing innovative farm and food policies to generate a competitive environment.

THE DUALITY OF THE AGRIFOOD SYSTEM

A peculiar feature of the Agrifood changes in Italy is the regional specialization. The performance of the food industry depends quite heavily on local agricultural production. This interdependence is due to the influence of specific geographic and climatic factors. An input-output matrix from 1975 shows that food industries derived 61% of the inputs from agriculture, 14% from imports, 16% from re-use of domestic food industry products and 5% from re-use of imported food (Bernin Carri, 1990; Gios a.o., 1982; Terrasi, 1985).

Agrifood districts in Italy are the evidence of local concentration of technologies and resources to accomplish competitiveness of food enterprises. There is a system of local enterprises usually of small scale and with flexible technologies, which are able to compete in the subsectors on which they are focusing. They significantly affected the performance of the agrifood system in different regions, stimulating the increase in occupation and value added. Agro-industrial districts are spreading in the specialized areas of the northern regions for poultry, fresh and preserved vegetables, meat packing, dairy products, wine and in some southern areas.

It appears that the interaction between farmers and private or public institutions, which assist farm and food enterprises, are key-factors in making agriculture competitive.

Socio-economic and geographic conditions have influenced the patterns of development. The ratio of VAFC of the food industry in 1989 to VAFC of agriculture was 75.3% in Northern regions, 51.4% in the center and only 23.6% in the south. During the period 1980-88, the VAFC of agriculture increased by 81% in south and 86% in the north, despite the production restriction imposed on milk and cereals. The south represents 42% of the total Agrifood value added and the long term projections indicate a further decrease. The duality of Italian agrifood system appears from the development in the ratio between the VAFC of the agriculture and the total GDP, which in the period 1980-88 passed in the south from 11% to 7% and in the north-center from 5% to 3% (see Table 2). In the same period, the productivity of agriculture measured in per-capita VAFC at constant 1980 prices, increased by 27% in south from an initial value of 6.774 (mio lire), in north-center increased

TABLE 2. Main indicators of the agrifood sector in Italy. Period 1980-88

Name		South		North-Center		Italy	
		1980	1988	1980	1988	1980	1988
(1) VAFC Prim.	bil	10016	18148	13577	25309	23593	43457
(2) VAFC Food	bil	2101	4797	7140	18999	9240	23796
(3) VAFC Total	bil	91698	255663	289946	788770	38164	67253
(4) WU Prim*	000	1479	1186	515	1321	2994	2508
(5) WU Food*	000	114	84	330	304	444	388
(6) WU Total	000	1593	1270	845	1625	3438	2896

Source: INEA and other sources
* data referred to 1987
VAFC = Value added factor cost
WU = Total working unit
Source: Our elaboration of Istat data

by 29% from an initial value of 8.960 (mio lire). In 1988 the VAFC in south was 4% less in respect to the north in 1980 and 34% less in 1988. The growth patterns are quite similar but differences in productivity are still consistent and persistent as suggested by the VAFC ratio between south and north-center passing from .74 to .72 in the period 1980-1988.

CONSUMER BEHAVIOR

The demand for food is subject to the influence of many factors: whereas food expenditure declined as a percentage of total private expenditure, it is growing in absolute terms and directed to products with higher value added. Consumer choice has been greatly influenced by the globalization of food markets, demographic changes, ageing, presence of multi-ethnical groups. Table 3 shows that consumer food expenditure has been growing during the period 1970-1989, at a lower rate than total private consumption expenditure (total) and than per-capita income: the ratio between the expenditure on food, beverage and tobacco and the total private expenditure reduced from an initial 39% to 22%; the value is close to the European average. A growing share of food expenditure is spent on high value added foods: processed food and food consumed away from home (afh). It reflects the changes in social habits and in income.

It is estimated that in 1989, 12 million meals were consumed every day away from home, representing 48.5 thousand billion lires on an annual basis. It increased another 5% in 1990. Geographic and regional differences are relevant: while national per capita monthly expenditure for food and drinks consumed out of home was estimated 215,000 lire, it amounted to 224,000 in the north, 228,000 lire in the center (influenced by the presence of the capital city) and 197,000 lire in the south. Food expenditures are highest for meat (28% of total expenditure) followed by fruit and vegetables (22%), dairy products and eggs (14%), cereal-based products (11.7%), fish (6%), wine and alcoholic beverages (5.5%).

Italians have a 'Mediterranean diet' of fresh products, oil, pasta, fresh fruits and vegetables and other typical foods. They are mixed

TABLE 3. Consumption expenditure in Italy (billion lire; per-cap inc. x 1000)

Year	FBT[1] consumption Abs. val.	ind.	Priv.final cons. exp. Abs val.	ind.	Ratio FBT/Tot	Per cap. Abs val.	Income ind.
70	15562	100.0	39992	100.0	.39	1596	100.0
75	29328	188.5	85972	215.0	.34	3459	216.7
80	67658	434.8	236603	591.6	.29	8742	547.7
85	127531	819.5	499587	1249.2	.26	17727	1110.7
89	160023	1028.3	732966	1832.6	.22	24213	1517.1

Source: Annuario Statistico Italiano and UN-National Accounts.

[1] FBT = Food, Beverages, Tobacco

with a growing quantity of imported foods that represent the internationalization of consumer tastes.

THE INTERNATIONALIZATION OF AGRIFOOD MARKETS

Important determinants of the internationalization of the agrifood markets are the growing interdependencies between the markets of different countries (Barkema a.o., 1991; Porter, 1990). One important factor is represented by the growing influence of foreign MNC's that operate in Italy and Italian MNCs: Ferruzzi, Barilla, Arrigoni, Parmalat operate in world markets and export substantial quantities of their production.

The domestic supply of agrifood products is not enough to satisfy domestic demand for food. This is certainly an important reason for the penetration of MNC's in domestic markets but other factors like inefficiency, lower barriers to entry and aggressive financial and commercial strategies of MNCs contribute to their consistent penetration in domestic markets too (Corsani, 1988; Galizzi a.o., 1989; Linda, 1988; Mariani, 1990).

Table 4 reports the evolution of the trade balance for the last thirty years with special reference to Italian dependence on international trade.

Italian food import surplus was quite impressive in the period 1970-1980. There is in particular a deficit in meat and dairy products.

The propensity to import increased at a lower rate and propensity to export at a higher rate in the period 1980-1990 as compared to the period 1970-1980.

THE FOOD INDUSTRY

In the first part of the eighties, average turnover of the food industry in constant prices, progressed at an annual rate of 2%, in comparison to a growth of 1% of manufacturing, and even accelerated to 3.5% in the second half of this period. At constant prices, the VAFC of the food industry increased during the same period annu-

TABLE 4. Trade balance of Agrifood products. (billion lire, price 1990)

Macroeconomic Aggregates	1970	1980	1990
Total Agrifood production (1)	81044	87228	87436
Import	19631	28332	30268
Export	6593	10250	13455
Balance (exp-imp)	-13038	-18082	-16813
Volume of Trade (2)	26224	38582	43723
Apparent Consumption (3)	94082	105309	104249
Indicators:			
Degree of self sufficiency (4)	86.1	82.8	83.9
Propensity to import (5)	20.9	26.9	29.0
Propensity to export (6)	8.1	11.8	15.4
Degree of business coverage (7)	33.6	36.2	44.4

(1) Value of Agriculture plus Value added of the Food Industry.
(2) Total import plus export
(3) Agrifood production plus import minus export
(4) Ratio production/Consumption
(5) Ratio import/consumption
(6) Ratio export/production
(7) Ratio export/import

Source: INEA, L'agricoltura conta

ally by 2.5%. Food industry was in the period the third contributor (10.8%) to GDP, after the machinery industry, 28.9%, and chemical industry, 17.2%.

It contributed in 1989 2.2% of the contribution to GDP and 1.6% to domestic employment. Growth differed consistently between branches of the food industry: most expansive were sugar and mineral water. The latter increased its turnover by 77.1% in the period 1985-90; it is one of the most active sectors in M&A (merger and acquisitions).

The First Processing of Agricultural Products

The industry, intermediate between the agricultural sector and the food industry, is defined by ISTAT: "Processing activities performed by the agricultural sector." In 1981 (Census data) this sector comprised more than 17 thousand enterprises and employed almost 100 thousand people. These enterprises usually perform a first transformation, packaging, conservation or refrigeration of the agricultural products.

In many agrifood districts, many processing activities are performed by co-operative establishments that operate close to the member's farms which deliver the products. They facilitate the transition from agriculture to agrifood system by integration of the functions performed at different levels of the food chain, developing food markets, solving some of the fiscal-financial and organizational problems. Their size, in terms of employment are higher (11 units on average), as compared to the average of the food industry (8 units), probably due to the labor legislation.

These co-operative establishment have an important share of the value realized in primary sector: 63% for butter, 47% for cheese, 43% for fruits, 37% for frozen products, 40% for wine. This type of enterprise gave in Emilia Romagna, the most important region for the turnover of the food industry (70% realized by co-operatives), the largest contribution to the organization of the agrifood sector. There are two organizational models of major co-operative groups: Confco-operative and Lega. Confco-operative is a federative model structured in five federations that represent the major activities performed in the agrifood subsectors. The system is working with three levels of co-operatives: primary, second and third degree co-operative organizations specialized to coordinate the collection, processing and marketing functions at local, regional and national-international levels.

The independence of each subsector seems to facilitate the specialization of product lines preserving the autonomy of members but makes it more difficult to develop multiproduct cooperative activities and collaborations for commercial strategies and research and development projects.

The other group, LEGA is based on a Holding organization that has the objective to centrally co-ordinate all the activities of the agrifood sector performed by the affiliated co-operatives operating in all agrifood subsectors. In this case the problem seems to be the opposite: the difficulty to supervise and control all these activities with a competent management and control the political influence.

CHANGES IN THE FOOD INDUSTRY

The food industry (NACE 41/42 or ISIC 311..314), like the manufacturing industry, started to grow rapidly during the sixties as a result of innovations in food technologies and in food products, that stimulated the demand for food.

The evolution of the food industry in the past is described in Table 5 reporting Census data for the entire industry in the years 1961, 1971, 1981. Total industry is compared with the companies employing twenty or more persons. The latter is considered more representative for the industry, with respect to the relevant structural changes.

The indexes show that structural changes are not very relevant in these years: the average of eight persons engaged per unit for the entire industry remained almost unchanged over the entire period; also for the industry with 20 or more persons per unit the decline of two units in 20 years doesn't represent a significative change.

In the first decade, the concentration is shown by the decline of the total number of companies and the increase of those with 20 or more persons engaged; in the second decade the situation reversed: structural indicators indicated a situation becoming closer to that of 1961, with the decline of larger units to 5% of the total and representing 57% of the persons engaged. For the following years, the

TABLE 5. Comparison between the large and the entire food industry

Year	Number of Employees				Number of Ratios		
	Total (1)	>=20 (2)	Total (3)	>=20 (4)	Ratio (3)/(1)	Ratio (4)/(2)	Ratio (2)/(1)	Ratio (4)/(3)
61	56640	2757	423508	243187	7,48	88,21	0,05	0,57
71	49032	2886	400699	241828	8,17	83,79	0,06	0,60
81	53006	2774	418568	238711	7,90	86,05	0,05	0,57

Source: Elaboration of Census data published by Istat.

data published by Eurostat (Structure and activity of industry: annual inquiry) indicate that the number of enterprises with twenty or more persons engaged increased by 10%, while the number of persons engaged remained quite unchanged.

MERGER AND ACQUISITION: STRATEGIES TO CONTROL THE FOOD MARKET

The recent history of the food industry is characterized by a substantial number of mergers and acquisitions, most of them majority acquisitions (Linda, 1988; Loseby a.o., 1992; Ravazzoni, 1991). BSN the French group through IFIL acquired the control of some important Italian groups like SCIA (87), Star (89), Peroni (88), the greater producers of pasta Agnesi and Ghigi and other smaller producers. Nestlé acquired the control in 1988 of Buitoni group and "King's prosciutti" a prestigious domestic label. In the alcoholic beverage sector, Artois-Piedboeuf acquired in 1986 36% of Von Wunster, Dreher-Heineken and in 1989 Moretti and Prinz were acquired by the Canadian group John Labatt. In the wine sector, Seagram acquired the control of Maschio group and Barbieri, Pernod-Ricard acquired Canei. On their turn Italian groups penetrated in foreign markets, like the Cesare Fiorucci (ham and salami), Lavazza and Segafredo-Zanetti in the coffee sector, Ferrero for chocolate, the establishment in the USA of a pasta establishment by Finpetrini, the expansion in the UK and Spain of the Consorzio Conserve Italia, the constitution of Multitrade in Spain and Beca Greece by Beca-Italy, the second largest company in meat processing. Martini e Rossi acquired the control of Calvados Boulard Benedectine from Remy Martin and realized a joint-venture with the British Boss and American Bacardi for the distribution of its products on these markets. Ferruzzi, the best performing agribusiness group after Eridania and Beghin Say acquired in this period CICA (Brasil), Guadalco (Spain) and the European industrial divisions of American companies CPC (now Cerestar) Central Soya, Lesieur (French group) and Koipe (Spain) leader in oil production (Galizzi a.o., 1989).

Of the 271 M and A operations, recorded in the period 1986-89, 141 (52%) were majority–or control acquisitions–42 (15%) were minority acquisitions, 15 (6%) were mergers, 68 (25%) were joint ventures and commercial agreements and 5 (2%) other operations (Mariani, 1990). The tendency to establish a form of "non equity" partnerships is to avoid 'risk capital' operations, that imply for smaller enterprises, in case of failure, the risk to lose autonomy or disappearance from the market (Ravazzoni, 1991; Swinbank a.o., 1983).

Important objectives achieved with this collaboration were:

i.   the uni- or bilateral transfer of technological and commercial know-how,
ii. the exchange of goods and services,
iii. interfirm collaboration in one or more functional areas of the agribusiness. The largest corporations (with turnover greater than 100 million $) were oriented to reach dominant market positions by pursuing strategic objectives like the control of the production and distribution of leading products and the acquisition of important labels. In this case there was a preference for joint ventures in which partners got involved in 'risk sharing' operations.

The ten largest groups listed in Table 6 realized more than 50% of the majority or minority acquisitions and joint ventures.

The joint venture of Galbani Alivar, Barilla and Ferrero indicates that the advantages for participants are mainly scale economies, the creation of barriers against competitors and better performance of oligopolistic strategies to control markets.

Merger and acquisitions differ between industry and markets. The greatest number of acquisitions (majority share and mergers) was realized in the wine sector (10%), followed by bread and bakery (7.1%), dairy (7.1%), pasta and rice (6.4%), mineral water (4.3%), fruit and vegetables (4.3%). Almost two thirds of the majority acquisitions of the last three years in the food industry had the objective to increase the control of the market. When acquisitions were made in areas not correlated to the "core business," the goal was to spread risk over different activities, to split profits in order to reduce taxation, to advance in vertical integration, to acquire the control of different enterprises in a business line. Recent events show a number of horizontal strategies oriented to achieve a better

TABLE 6. Distribution of turnover and position of the main Groups in Italian Food Industry

Sector of activity	Market value bill. lire	Principal enterprises or groups	Market quota
Pasta	3.000	Barilla	32
		Nestlè	6
		BSN-IFIL	6
Ice cream	1.200	Unilever	40
		SME	29
		Sammontana	10
Frozen product	1.700	Unilever	52
		SME	20
		Sammontana	9
Process. Fruits and Vegetable	1.800	SME	20
		Cons. Italia	15
		Fedital	10
Manufacture of vegetable oil	3.200	Unilever	11
		Ferruzzi	10
		SME	5
Bakery products	4.000	Barilla	26
		SME	11
		BSN-Saiwa	5

Product	Sales	Companies	Share
Dairy products	9.500	BSN-Ifil	13.5
		Kraft	5.5
		Parmalat	4.5
Processed meat (salami)	7.000	Nestlè	5
		BSN-Ifil	4
		Fiorucci	4
Mineral water	1.700	BSN-Ifil	16
		Ciarrapico	11
		San Pellegrino	9
Beer	1.100	BSN-Ifil	40
		Heineken	30
		Prinz	9
Coffee	2.200	Lavazza	22
		Procter & G.	8
		Nestlè	5
Chocolate, sugar and derived	2.200	Ferrero	20
	Nestlè	Ferrero 16	

Source. Our elaboration from different sources

control of the core business areas. In the sectors mineral water, beer, bread, bakery, pasta, slaughtering and salami, 100% of the acquisitions were realized in the same sector. The global turnover of the largest 53 Italian food groups was estimated in 1989, 20 billion dollars corresponding to 25% of the total consolidated turnover of the Italian food industry and 75% of Nestlé (Ravazzoni, 1991). The 60,000 persons engaged in this group of enterprises, represent 15.4% of total employment in the food industry. At least 40 of these enterprises participate in foreign groups and in 18 of these enterprises the total or the majority of the assets, is controlled by a foreign group. Forty-four percent of the turnover of the first 10 major companies in Italian Food Industry has been covered by the first three groups which are controlled by Unilever, BSN and Ferrero.

Table 6 indicates, for 12 food sectors the aggregate business (market value) realized by major groups and their market. There is only one sector, the processed fruit and vegetable where the three major groups are Italian; at the opposite beer, coffee and chocolate are controlled by foreign groups.

The data show consistently a substantial financial control by foreign groups in the most important sectors of the Italian food industry.

We list three different types of acquisitions:

i. Large groups that made acquisitions in the business area, being their core business. Cases in point are Ferruzzi (cereal and oil), Barilla (pasta), BSN-Ifil (dairy products);
ii. Foreign groups that invested in specialized activities, where Italian groups had a substantial market control. (BSN-Gervais Danone, Nestlé, Dart & Kraft);
iii. Smaller enterprises that use the acquisitions to enter and control the market they were interested in. (Itafin, Can-Fin, San Carlo Gruppo Alimentare).

At the end of 1988, 15 of the leading 42 corporations were controlled by MNC's. The interest of foreign groups in Italian food industry can be explained by four arguments:

i. the limited barriers to entry due to the prevailing financial weakness of the domestic firms; for some of these, after the acquisition they evolved towards a subsidiary unit of the MNC, with most of the managerial functions transferred to the direction of the MNC.

ii.   the synergetic effect of the "niche-label" derived from the position in the domestic market, combined with the commercial power of the label of the foreign Group.

iii. the possibility to exploit a quite traditional market with innovative products that are accepted by Italian consumers.

iv.   the need to gain scale economies to justify the growing expenses in research and development for new products.

CONCLUSIONS

The Italian agrifood system is evolving under pressure of international competition that brings about structural adjustments to new conditions of competition.

Changes in traditional agriculture are required with market transformations, international competition and the McSharry reform which indicates the conversion of the CAP from assisted agriculture towards rural development.

The pressure of the United States to abolish the present CAP system of aid and preferences, which alters competitive conditions, is another important factor of structural adjustment.

Despite the understable resistance of many farmers to the agricultural reforms the evolution toward a competitive agrifood system continues because single states are increasingly unable to afford the growing costs of agricultural surpluses. Whereas the future single and open Common Market will not produce dramatic effects or more conflicts as compared to the present situation, some changes are expected from the removal of the remaining trade barriers, the unified system of marketing operations, rules of pricing, negotiations and transportation standards. It will push further the development of multinational groups, which will gain competitive advantages in economics of scale, marketing intelligence, risk diversi- fication and management co-ordination.

The Maastricht treaty will probably have financial consequences for the regulation of exchange rates and the eventual introduction of a common currency which will facilitate financial transactions. The regulation of new institutions, which are becoming more common, like option trade and futures markets contribute to the efficiency of food markets. Changes of the agrifood system are accelerated by the growing influence of greater mobility and the influence of media and communication on consumer behavior. Weaknesses of Italian agriculture are the farm size inferior to most other EEC countries and poor competition on costs, quality and standardization as required by processing industry and final markets. However, this statement must be differentiated according to the regions. The northern regions are in general better organized. The Agrifood districts are the evidence. They concentrate resources and knowledge to create conditions for an excellent market performance. In these regions employment in agriculture is on average 4% and the size of farms is greater than 10 hectares. Structural adjustments observed can be like the exit of farmers and agricultural laborers from the farm sector, the diffusion of part-time farming, the concentration of farms and their co-ordination with the food processing industry facilitated by the presence of a cooperative system.

In the south structural adjustment is more difficult for a number of reasons. The most important obstacle seems to be the allocation of the redundant labor force outside the farming sector. Nevertheless the situation varies greatly by region and within regions: in Puglia, Campania or Sicily there are also some specialized areas for the production of vegetables, wine, citrus products and oil.

The performance of the food industry is affected by the inefficiencies of agriculture, the organization of the distributive sector and the specific attitudes of Italian consumers for *home made* foods. The expansion of many Italian companies abroad has demonstrated the opportunities to expand in foreign markets. Their limits can be summarized in two points:

i.   They tend to concentrate their strategies abroad in traditional products like pasta, tomatoes, vegetables, where competition is high. The acquisitions of important Italian groups by foreign MNCs demonstrate that this sector is no longer dominated only by large Italian companies.

ii. They tend to enter into niche markets that are more appropriate for Italian foods. This sector requires strategies based on quality standards, image building, customer services, pricing strategies based on experience and skills.

More market penetration can be achieved by merger & acquisition and new market opportunities can be created by participation in foreign groups that control substantial distribution and perform successful marketing strategies.

REFERENCES

Balestrieri, G. (1988). "Aspetti dimensionali e di forma giuridica delle imprese nella localizzazione dell'industria alimentare in Italia." Atti del XXVI Convegno SIDEA, Strategie ed adattamenti nel sistema Agroalimentare, Il Mulino, Bologna.

Barkema, A., Drabenstott, M, Welch, K. (1991). "The quiet revolution in the U.S. Food Market." Economic Review, May-June.

Bernin, Carri C. (1990). "L'analisi linkage applicata matrice Italiana e le branche agricolo-alimentari." Riv Econ. Agraria, 42, pp. 257-288.

Bertelè, U. (1988). "L'agricoltura Italiana nel contesto interno ed internazionale," Riv. Pol. Agraria, 1.

Bertelè, U. (1989). "Industria alimentare Italiana e mercato mondiale," Unavicoltura, 11, pp. 22-35.

Bagarani, M., Magni C., Mellano, M. (1988). "La specializzazione Agricolo-alimentare nelle regioni Italiane nei diversi gradi di integrazione," Atti del XXIV Convegno SIDEA, Strategie ed adattamenti nel sistema agricolo-alimentare, Il Mulino, Bologna, pp. 155-177.

Boucher, M. ed. (1989). "Strategies Industrielles Mondiales," Cahiers Francais n. 243, oct-dic.

Cannata, G. (1988). "Risultati di un'indagine su alcuni comparti dell'Industria alimentare in Italia." Quaderni Istituto di Studi Economici, Libera Università degli studi sociali, Roma, 36.

Chang Ting Fa, M. (1987). "Le trasformazioni economiche nei paesi della CEE ed il ruolo dell'Agribusiness," Studi di Economia e Diritto, 2.

Corsani, A. (1988). "Les Industries Alimentaires Italiennes, Allemandes et Francaises depuis les années 50: Convergences et Differences." Revue d'Economie Industrielle, 44, 2ème trimestre.

Fabiani, G. (1984). "La collocazione internazionale dell'Economia Agroalimentare Italiana," Polit. Economia, 15 pp. 53-58.

Fanfani, R., Montresor, E. (1991). "Filiere Multinazionali e Dimensione Spaziale nel Sistema Agro-Alimentare Italiano." Quest. Agraria, 41, pp. 166-201.

Fanfani, R., Gatti, S. (1992). "Productivité et competitivité dans l'Industrie Agroalimentaire italienne," Economie Rurale, 207, pp. 17-25.

Frigero, P. C. (1978). "Analisi della produttività della Industria Agroalimentare Italiana e problemi di ristrutturazione del settore," Riv. Econ. e Polit. Industriale, 4, pp. 175-222.

Galizzi, G. (1981). "Sistema Agro-alimentare e linee di Politica Agraria." Riv. Econ. Agraria, 36, pp. 55-84.

Galizzi, G. (1987). "Integrazione Verticale in Agricoltura: Meccanismi ed Aspetti Istituzionali." Riv. Pol. Agraria 2.

Galizzi, G., Linda, R. a cura di (1989). "Strategie di internazionalizzazione dell'industria alimentare europea," Rivi-sta milanese di economia, serie quaderni, n° 18, ott-dic.

Giacomini & others (1991). "Il sistema Agro-alimentare nella struttura dell'Eonomia Italiana." F. Angeli ed..

Gios, G., Miglierina, C. (1982). "L'evoluzione del sistema agro-alimentare comunitario: Un'analisi condotta sulla base delle tavole Input-Output," Riv. Economia Agraria, 37, pp. 955-976.

Hazledine, T. (1991). "Productivity in Canadian Food and Beverage Industries: An interpretive Survey of Methods and Results." Can. Journ. of Agric. Economics, 39, pp. 1-34.

Iacoponi, L. (1990). "Distretto industriale Marshalliano e forme di organizzazione delle imprese in Agricoltura," Riv. Econ. Agraria, 4.

Ievoli, C. (1986). "Economie di dimensione e struttura di mercato nell'Industria alimentare," Quest. Agraria, 23, pp. 95-125.

INEA (1991). "L'Agricoltura Italiana in cifre."

Linda, R. (1988). "Crescita, diversificazione e denominazione nelle strategie delle mega-aziende europee dell'Industria alimentare e delle bevande." Riv. Pol. Agraria, 4, pp. 37-55.

Loseby, M., Matteucci. B. (1992). "L'evoluzione recente della struttura dell'industria alimentare Italiana in vista del completamento del mercato unico." Riv. Pol. Agr., 2.

Magni, C. (1982). "Il sistema agroalimentare," Quest. Agr., 6.

Mariani, A. a cura di (1990). "La struttura dell'industria alimentare Italiana," F. Angeli ed.

Porter, M. E. (1990). "The competitive advantage of Nations," New York, Free Press.

Ravazzoni, R. (1991). "Concentrazione ed internazionalizzazione dell'Industria Alimentare Italiana." Riv. Pol. Agr., 1, pp. 29, 54.

Rosa, F., Galizzi, G. (1991). "A comparative analysis of Macroeconomic changes for the OCDE Countries and the role of the Agrifood sector." in Food Marketing and Food Industries in the Single European Market, ed. by H. E Buchholz and H. Wendt, Braunschweig-Volkenrode (FAL).

Terrasi, M. (1985). "I fattori di localizzazione dell'Industria Alimentare." Riv. Econ. Agr., 1.

Swinbank, A., McInerney, J., Burns, J. (1983). "The Food Industry. Economics and Policies," Published with the Commonwealth Agricultural Bureaux.

Venturini, L. (1991). "Vertical integration and Deconcentration: An analysis of cross-country structural changes in the European Frozen Food Industry." in Food Marketing and Food Industries in the Single European Market, ed. by H.E Buchholz and H. Wendt, Braunschweig-Volkenrode (FAL).

Agricultural and Food Marketing in the United Kingdom: A Review of Some Current Issues

Jeffrey Lamont
Christopher Ritson

SUMMARY. Given the diversity of agrifood marketing in the U.K., in terms of the range of agricultural commodities and value-added products produced, the many and varied marketing structures used, and the numerous legislative and policy instruments involved, it is not possible in this paper to provide a comprehensive coverage of all aspects of the system. The approach has been rather to provide the volume's international readership with a selective view of the U.K. agrifood marketing system, concentrating on some topical issues. This approach has the benefit of allowing more in-depth consideration of important issues, and a focus on topics where the U.K. exhibits a degree of uniqueness amongst its E.C. partners. The paper therefore presents an international readership with an analysis of truly "*U.K. issues.*"

The first section highlights the impact of two major factors in the *marketing environment* upon the U.K. system;

Jeffrey Lamont is Marketing Lecturer, Department of Agricultural Economics and Food Marketing, University of Newcastle Upon Tyne, England. Christopher Ritson is Head of the Department of Agricultural Economics and Food Marketing, University of Newcastle Upon Tyne, England.

[Haworth co-indexing entry note]: "Agricultural and Food Marketing in the United Kingdom: A Review of Some Current Issues." Lamont, Jeffrey, and Christopher Ritson. Co-published simultaneously in the *Journal of International Food & Agribusiness Marketing* (The Haworth Press, Inc.) Vol. 5, No. 3/4, 1993, pp. 83-112; and: *Food and Agribusiness Marketing in Europe* (ed: Matthew Meulenberg), The Haworth Press, Inc., 1993, pp. 83-112. Multiple copies of this article/chapter may be purchased from The Haworth Document Delivery Center [1-800-3-HAWORTH; 9:00 a.m. - 5:00 p.m. (EST)].

a. trends in U.K. consumer behaviour with respect to agrifood products, and
b. the effect of "1992" on the U.K. agrifood sector.

The remainder of the paper reviews trends within the *market structure* of the UK system, concentrating upon three issues;

a. structural concentration and changes in distribution in the U.K. food marketing sector;
b. the role and future of the U.K. agricultural commodity marketing boards; and
c. the role and status of co-operatives in the U.K. agrifood marketing system.

INTRODUCTION

The British agricultural marketing system exhibits a high degree of diversity. Not only is there a wide range of agricultural commodities produced in the U.K., but the structures and systems used for marketing these commodities are many and various. There exists a combination of long standing traditional marketing practice, compulsorily organised marketing for some products, voluntary co-operative marketing of several products, and for others a highly complex pattern from the farm gate through to the consumer. It is not within the scope of this article to discuss the marketing of all the agricultural commodities produced in the U.K. Such a discussion would be very superficial and would fail to provide sufficient detail on the most important topical issues facing agricultural marketing in the U.K. today; issues which will be of greater interest to the volume's international readership. A more wide ranging discussion of the agricultural marketing system in the U.K., including the marketing of specific commodities, legislative and policy aspects of agricultural marketing and the utilisation of different marketing channels and techniques by U.K. agri-marketers, is provided by *Barker, J. (1989)*.

The purpose of this article is, rather, to provide an international readership with a more selective view of the present U.K. agrifood marketing system, by focusing on some topical issues. This approach allows more in-depth analysis of specific marketing issues, and it facilitates a focus on those topics where the U.K. exhibits a degree of uniqueness amongst its E.C. partners. The article therefore presents an international readership with a discussion of truly *U.K. issues.*

The title chosen for the article (Agricultural and Food Marketing) reflects the authors' belief that the marketing of agricultural products and the marketing of food cannot usefully be discussed independently. The two are obviously interdependent, with trends in food marketing having a major impact on agricultural marketing (and vice-versa). In 1990, U.K. consumers spent £31.7 billion on food products, of which £19 billion (60%) was on U.K. produced agricultural products; *Wilson, J. (1991).* The scope of this article therefore encompasses not only the marketing of agricultural commodities in the U.K., but also the marketing of value-added food products. It is concerned with the U.K.'s total *agrifood marketing system.*

Despite the selective approach adopted, the authors wish to maintain a structure which is consistent with that of other articles in this special edition, in order to facilitate comparisons between countries. Section 2 of the article thus concentrates on issues concerned with the present *marketing environment* facing the U.K. agrifood system, and Section 3 discusses currently important topics relating to the *structure* of the U.K. system.

Due to the interdependence of the U.K. agricultural and food sectors, the food consumption and purchasing habits of the U.K. represent a major environmental variable impacting upon the marketing of agrifood products. The dynamic nature of consumer behaviour is one of the major factors forcing changes to the U.K. marketing system, and a large part of Section 2 is devoted to describing the effects of the *consumer behaviour variable* upon agrifood marketing practice. *Political and policy variables* in the marketing environment are also bringing about major changes in the agrifood system, and the remainder of Section 2 will discuss E.C. legislation as it affects the marketing of U.K. agrifood products.

Section 3 will review three major topical issues relating to the structure of the U.K. agrifood marketing system. First, the degree of *structural concentration* and accompanying *changes in distribution* which are occurring will be discussed. Second, the debate concerning the future of the *commodity marketing boards* in the U.K. will be explored. Finally, the role and status of *agricultural marketing co-operatives* within the U.K. agrifood system will be considered.

THE MARKETING ENVIRONMENT

(a) General Trends in Consumer Behaviour with Respect to Agrifood Products

The traditional view of the British food consumer is of someone who will settle happily for "meat and two veg" followed by apple pie and custard, and a strong cup of tea with two sugars. However, this view is increasingly an anachronistic one, as the past decade has seen revolutionary changes in the patterns of U.K. food consumption. Indeed, *Ritson and Hutchins (1990)* have coined the phrase "the consumption revolution" to describe the intensity and rapidity of this change in consumption patterns. These writers also suggest that, if anything, the pace of this change is now accelerating. As *Wilson, J. (1991)* points out, the traditional consumer postulated above is increasingly being replaced by a much more progressive and cosmopolitan consumer, who is more likely to demand Chicken Kiev, baby corn, fresh fruit salad with fromage frais (all in environmentally-friendly, recyclable packaging), washed down with mineral water. A revolutionary change indeed!

The central questions facing the U.K. agrifood marketing industry in this respect are, first, what are the underlying causes of changes in consumption patterns; second, what are the specifics of the "consumption revolution" in terms of changes in agrifood product preferences; and third, what impact do these changes have upon the providers of agrifood products? *Ritson and Hutchins (1990), Wheelock (1990) (1986), Wilson (1991), Jamieson (1988), Thompson (1988), Wright (1988), Mackenzie (1990)* and *Paulson-Box and Williamson (1990)* all provide useful discussions of these three central questions.

Ritson and Hutchins (1990) argue that, unlike earlier periods of rapid development in U.K. food consumption, the major trends now

emerging are primarily driven by the attitudinal and social behaviour characteristics of U.K. consumers. To illustrate this point,
Figure 1 provides a simplified interpretation of the factors which
have influenced changing patterns of food consumption in the U.K.
during the life of the National Food Survey.

During the "wartime austerity and rationing" period, individual
food choice was largely "imposed" by availability. During the
"return to normal diets" phase, the end of wartime rationing and
more plentiful supplies of food allowed British households to return
to what would then have been regarded as "normal" diets. (The
"traditional" consumer postulated at the beginning of this section
was prevalent during this period.) From the mid 1950s through to

FIGURE 1. Factors Influencing Changes in U.K. Food Consumption Patterns, 1940-1990

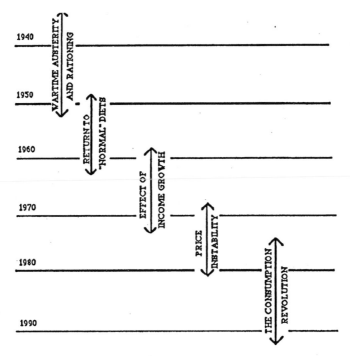

Source: *Ritson and Hutchins (1990)*

the 1970s, the U.K. experienced a general rise in living standards, and this became the major determinant of changes in food consumption patterns. Increasing affluence allowed food consumers selectively to increase their consumption of "preferred foods," whilst at the same time selectively "de-listing" other foods (known as "inferior goods"). Table 1 illustrates foods which can be classified as either "preferred" or "inferior" products during the twenty year period 1955-1975.

From the mid-1970s to 1980, prices became much more important as a determinant of food consumption patterns. *Ritson and Hutchins (1990)* suggest several reasons for this, including the world commodities price boom; the adoption of the Common Agricultural Policy; the U.K. drought of 1975/76; and the food subsidy programme. All of these factors combined to make the last five years of the 1970s a period marked by extreme volatility in the retail prices of many food products in the U.K.

And so to the 1980s, the decade of the "consumption revolu-

Table 1: Illustrating Food Products for which the Period of Income Growth (1955-1975)

Caused:

a) a Rise in Average Levels of Consumption ("Preferred" Products);

Cheese	Beef	Salad Vegetables	Fresh Fruit	Rice
Canned Salmon	Pork	Salad Oils	Chocolate Biscuits	Coffee
Shell Fish	Chicken	Frozen Vegetables	Brown & W'Meal Bread	Ice Cream

b) a Fall in Average Levels of Consumption ("Inferior" Products).

Canned Meat	Margarine	Potatoes	Tea
Sausages	Lard	Dried Pulses	White Bread
Herrings	Canned Milk Puddings	Canned Vegetables	Oat Meal Products

Source: Ritson and Hutchins (1990)

tion." By 1980, prices were less volatile, world commodity price fluctuations had stabilized and U.K. food prices had absorbed fully the affects of adopting the Common Agricultural Policy. During the past decade the pace of income growth in the U.K. has slackened, and the strength of the relationship between income and consumption seems to have lessened. In determining what factors have become dominant in shaping food consumption during the 1980s, the U.K. National Food Survey provides useful data which make it possible to establish significant relationships between average consumption levels and changes in average prices and household incomes. These relationships can thus be used to estimate what proportion of changes in purchases can be attributed to price and income factors. What is left, the residual–known as the "underlying trend in demand"–must therefore be attributable to something else–probably fundamental changes in consumer tastes, attitudes and social behaviour. It should be emphasized that these are changes in underlying demand (i.e., demand-schedule changes mediated by attitudinal and social behaviour factors), not actual consumption; in some cases the figures may contradict actual sales trends, if sales have been strongly affected by positive or negative price developments. But the underlying trend in demand is a much better guide for food marketing than consumption changes–because it indicates a trend which may be expected to be sustained into a future of constant prices and incomes. This point is illustrated in Figure 2 for pork.

The diagram shows purchases (in weight per person), price (in real terms) and demand (purchases, with the effect of price and income changes removed), all converted to an index based on average values for 1977-1982 being equal to 100. There is a general perception in the U.K. that pork consumption, which has held up much better than the "more red" meats of beef and lamb has become more popular. But the analysis shows this not to be the case. "Demand" has been in decline; but purchases have been sustained by more competitive prices.(See Figure 3.)

During the 1970s, consumption reacted to price changes and demand was relatively stable; since the late 1970s however, price fluctuations are reduced and a strong underlying adverse trend in demand has set in pulling consumption down. For many food prod-

FIGURE 2

PORK

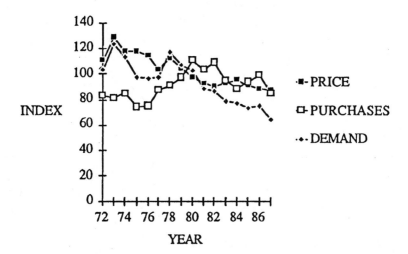

INDEX

■- PRICE

□- PURCHASES

·•· DEMAND

YEAR

ucts in the U.K. the period 1978-80 seems to represent a watershed–with strong positive or negative underlying trends in demand becoming apparent–the "Consumption Revolution."

Ritson and Hutchins, (1990) have fitted simple linear trends to the demand for over 150 food products or product groups and drawn up a "league table," ranking the products from those with the highest underlying trend in demand to those with the lowest. The "top ten" products are listed below:

1. "Other" fresh green vegetables (e.g., spinach, broccoli)
2. Coffee essences
3. Wholemeal and wholewheat bread
4. Frozen chips and other frozen convenience potato products
5. All "other" fats (e.g., low fat spreads)
6. Frozen convenience cereal foods (e.g., pastries and pizzas)
7. "Other" vegetable products (e.g., salads, coleslaw, pies, ready meals)
8. Fruit juices
9. Crisps and other potato products, not frozen
10. "Other" fresh fruit (e.g., melons, pineapples and exotics)

The bottom ten are:

1. Fresh white fish, unfilleted
2. Fresh peas
3. Processed fat fish unfilleted
4. Soft fresh, fruit, other than grapes
5. Instant potato
6. Offals, other than liver
7. Baby foods, canned and bottled
8. Canned peaches, pears and pineapples
9. Canned potatoes
10. Brussel sprouts

In providing a social behaviour/attitudinal explanation of these figures, *Ritson and Hutchins (1990)* point to the importance of two factors. First, a number of changes within the U.K. household–particularly the increase in the number of working women and the breakdown of traditional meal patterns in the home–have led to an

FIGURE 3. The typical U.K. demand trend is illustrated in FIGURE 3 for beef and veal.

BEEF AND VEAL

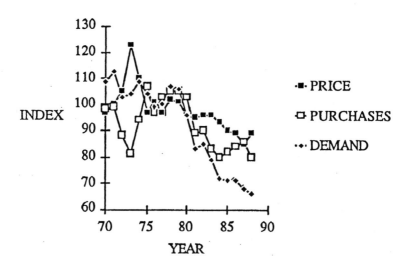

increasing demand for foods which are easy to prepare, quick to cook, and which come in individual portions. The "convenience food" culture is now well established in the U.K. This not only affects demand for the obvious foods (such as "frozen convenience" products) but also, for example, the form in which meat is presented (chops and mince, and ready-prepared cuts, rather than joints). A further indication of the extent to which the fast food culture has taken over within the U.K. household is provided by *Gofton and Marshall (1989)*. These writers report from a food diary study that 94% of meals involve less than 10 minutes preparation time, 51% involve no preparation time at all, 61% of all meals involve no cooking time, and only 7% involve more than 20 minutes cooking time.

The second factor identified by *Ritson and Hutchins (1990)* is what is known as the "vintage effect." The stage in the family life cycle–characterised in the National Food Survey by the recorded age of the housewife–provides one of the best explanations of differences in patterns of household consumption. The typical pattern is one of less than average consumption at early stages (young single people and young marrieds) rising to a peak when the housewife is in her fifties (the children have left the household and disposable income is high) and then declining towards average for pensioner households.

Consumption variations according to age can be attributed to two effects. First, the structure (size, number of children, etc.) of the household, proportion of meals eaten outside the home, and income of the household will be closely connected with the age of the housewife and this will influence patterns of consumption. Second, people form consumption habits as children and young adults, and carry these habits through with them as they grow older. The latter– the vintage effect–has a profound influence on which products display rising, and which declining, underlying trends in demand.

The British National Food Survey provides clear evidence of changing demand schedules for specific food commodities, and a social-behaviour/attitudinal explanation of underlying demand trends in terms of changing household structure and dynamics, the family-life-cycle "vintage" effect, and changing attitudes to healthier eating. In addition, other writers point to the increasing influence of travel abroad, international media communications, greater social mobility of the U.K. population, and concern for the environment and third world countries, upon U.K. food consumption patterns; *Wheelock, J.V. (1990) (1986), Wilson, J. (1991), Jamieson, M. R. (1988), Wright, G. (1988)* and *Mackenzie, D. (1990).*

In simple terms, the implications of the "consumer revolution" for those involved at the supply end of the U.K. agrifood marketing chain are clear. For the food retailers, the objective must increasingly be to provide consumers with appropriately convenient, healthy and increasingly exotic value-added food products. Manufacturers face the task of sourcing appropriate agricultural commodities for conversion to the required retail food products. Farmers, now more than ever before, must ensure that enterprise selection and farming methods are correlated with the tightly specified demands of the food manufacturing and retailing sectors. Increasingly exacting consumer-led demands are being exerted throughout the U.K. agrifood marketing chain.

A final aspect of changing U. K. food consumption patterns which, if not unique to the U.K., is certainly most pronounced there, relates to the impact of various "food scares." With trends toward convenience foods, (the so-called "menu" or prepared-meal dishes), have come many new product forms, new presentation technologies and preservation methods. Various writers have stressed how potentially health-threatening these new technologies can be; *Kerr et al, (1988); Lacey, R. (1989); Lacey, R. and Dealer, S. F. (1990); Gofton and Ness, (1991).* Over the past decade, major food scares have shaken public confidence in purchasing many food items, including milk, cheese, other dairy products, eggs, beef, and a range of cook-chill products. This has been caused by scares concerning Salmonella, Listeria, Alar, BSE, and the use of pesticides.

Gofton and Ness, (1991) argue that the actual evidence which forms the basis for such food scares seldom, in itself, generates a great deal of public response in terms of changing food consumption patterns. As *Beardsworth, (1990),* and *Gofton (1990),* argue, it is the effect of media coverage over time which has been the major factor in amplifying public perceptions of food risks, and ultimately causing changes in consumption patterns. This confirms the idea of *Cohen, S. (1970).*

The sometime devasting impact which food scares can have upon U.K. food markets (particularly in the case of the "Salmonella in eggs" scare), obviously represents a new and serious threat to the efforts of U.K. food marketers. This is particularly true when the effects of such scares are set alongside longer-term concerns over the relation between diet and the so-called "diseases of affluence"– cancer, stroke and coronary heart disease. In addition, a new anomaly presents itself for the U.K. food marketers, in that the pursuit of ever more convenient methods of food retailing, preparation and consumption, is often antithetical to the preservation of food purity and healthiness. This anomaly is comprehensively discussed by *Gofton and Ness, (1991)*.

(b) The "European Dimension": Its Impact on the U.K. Agrifood Marketing System

Although eighteen years of E.C. legislation have already impacted upon the U.K. agrifood system, of greatest topical interest is the impending completion of the "Single Market" after 31st December 1992. Nothing concentrates the mind quite like the imminence of a much-publicized event! However, the impact of this "event," upon the U.K. agrifood system is by no means easy to predict. *Paul Theroux (1989)*, reports that, when the Chinese Premier Chou En-lai was asked what he thought the significance of the French Revolution had been, he replied "It's too soon to tell."

The difficulty of forecasting the impact of "1992" upon the U.K. agrifood system is evident from the varied comments to be found in the literature. *Evanson (1990)* suggests that "as far as agricultural products, or the food industry are concerned, 1992 will be largely a non-event." In the same vein, *Peter Pooley*[1] *(1989)* an important British figure in Europe has said "I don't think 1992 is all that important for British Agriculture. We achieved our Common Market long ago and it just needs tidying up a bit." By way of contrast, *Dancey, R. J. and Jones, G. L. (1990)* warn that "whilst the final outcome of 1992 on individual U.K. agrifood sectors is difficult to predict, no agricultural or food-related business can afford to ignore

---

1. Then Deputy Director of Agriculture in the Commission of The European Communities.

its development." *Swinbank (1990)* echoes these warnings in respect of both U.K. consumers and U.K. agrifood businesses. *K. Galama (1990)* predicts "marketing wars" raging-between U.K. and "European" agrifood businesses, post-1992. Similarly, *Pickles (1990)* points to evidence that U.K. agrifood firms are not perceived to be a major threat by rival European firms, whereas the U.K. agrifood market is a heavily targeted market for firms from other European countries.

So, what should we make of these claims and diametrically-opposed counterclaims? Rather than err on the side of complacency, the authors believe that there does exist a sufficient body of evidence to suggest that completion of the single European market will have important implications for the U.K. agrifood marketing system. Whilst it is difficult to generalise across studies in the literature, three common areas of concern do emerge with respect to how "1992" will effect the U.K. system:

i. sector-specific effects will be felt; for example in the milk and livestock sectors;
ii. completion of the market will have profound implications for U.K. food and drinks manufacturing firms; and
iii. there will also be consequences for the U.K. agrifood consumer.

Implications for the U.K. Milk and Livestock sectors are well documented; *Dancey and Jones, (1990); Evanson, (1990); Gardner, (1989); Ritson and Swinbank, (1991); Wilkinson, (1984); Palmer, C. M. (1990)*.

The question of how the U.K. food and drinks manufacturing sector will be affected by "1992" is also well documented; *Pickles, L. (1990); Swinbank, A. (1990) and (1983); Galama, K. (1990);* and *Boakes, N. (1990)*. In general terms, the work of these authors can be summarized in a S.W.O.T. analysis of the U.K. food and drinks manufacturing industry vis-à-vis marketing in a single Europe:

Strengths The U.K. food industry is in a very strong position to benefit from trends to convenience and healthy eating in other European countries, because the U.K. convenience/healthy foods market is by far the most advanced in Europe.

Weaknesses Compared with their Continental counterparts, U.K. food manufacturers may be at a disadvantage due to a poor pan-European brand image. Few British brands are established European brands.

Opportunities Throughout Continental Europe, the major trends are the same: increased opportunities for prepared food products offering high quality, high nutritional value, freshness and convenience.

Threats U.K. food consumers will be increasingly offered an expanding range of products of Continental origin, as the U.K. food market is very heavily targeted by Continental firms as a major post-1992 opportunity.

In order to capitalise fully upon their major strength of being the most advanced convenience/health market in Europe, *Pickles (1990)* warns that U.K. food firms will need to ensure that they maximise potential cost economies. Pickles suggests that there are two types of U.K. food firm with particular differential advantages, which are most likely to succeed in the new single Europe. First, those with a strong niche position in the U.K. market, supplying high quality products economically to the major retailers, and second, those branded manufacturers able to expand beyond the U.K. However, in order to effect such expansion into a single Europe, U.K. firms must apply themselves to the major weakness of lacking a strong pan-European brand identity. European-wide branding is already evident in the products of many of the U.K.'s competitors, such as: pasta, e.g., Barilla, Buitoni; pasta sauces, e.g., Ragu, Dolmio; yoghurts, e.g., Chambourcy, Gervais Danone; dried sauces, e.g., Knorr.

The successful development of pan-European brands does however raise the question of the mutual acceptance of agrifood product standards by all E.C. member-states. *Swinbank, A. (1990)* provides a useful discussion of this issue in relation to the pan-Europeanisation of U.K. agrifood products. Traditionally, non-tariff trade barriers to agrifood products have been erected by many E.C. governments for very understandable reasons, such as the protection of their citizens against defective or dangerous products. The logic of the single market is that inter-E.C. differences in non-tariff trade barrier standards will be removed. Theoretically, therefore, it should no longer be necessary to produce products with different specifications for the separate E.C. national markets (although agri-food marketers may continue to do so, on the basis of distinct segmental differences). Economies of scale and mass production could therefore be reaped by U.K. agrifood firms in the pan-Europeanisation of their brands.

Conventional thinking on routes to achieving such equalising of E.C. product standards has always emphasised two legislative possibilities;

  i. the adoption of Community-wide legislation harmonising any offending features of divergent national agrifood product laws, and
  ii. action in the European Court seeking, under Article 30 of the Treaty of Rome, to declare invalid national import restrictions.

Early attempts at harmonisation led of course to considerable political resistance at the thought of "Euro-beer" or "Euro-bread" (much resented in the U.K.) at the expense of the diversity of national and regional products.

Subsequently, due to revision of the Commission's legislative program, the overriding principle of "mutual recognition" has evolved. As this principle is further validated by cases passing through the European Court, it is believed that increasingly member states will recognize the futility of defending protectionist agrifood policies before the Court, and will instead admit for sale on their own territory agrifood products which are

> lawfully and fairly manufactured and sold in any other member state, even if such products are manufactured on the basis of technical specifications different from those laid down by national laws in so far as the products in question protect in an equivalent fashion the legitimate interests involved. *Commission of the European Communities (1988)*

Given that one of the major tasks facing the U.K. agrifood sector is the successful development of strong pan-European brands, then the principle of mutual recognition, now well established, will facilitate the accomplishment of this task.

In terms of single-market effects upon the U.K. food consumer, *Swinbank (1990)* warns that whilst the consumer should benefit from a wider choice of quality food at lower prices, there may also be a disadvantage in the form of consumer confusion. He suggests that consumers could be faced with a bewildering range of quite different, but apparently comparable, products, causing a high degree of cognitive dissonance in consumers, and ultimate debasement of product value.

This, then, concludes this selective analysis of some of the major issues in the *marketing environment* facing the U.K. agrifood sector. Revolutionary changes in the food consumption behaviour of U.K. consumers is having a profound effect on the range and value-added specifications of demanded food products, and this poses major challenges to U.K. food retailers, manufacturers and farmers alike. Political considerations are also important, particularly in relation to "1992," where dramatic sector-specific effects are likely to be felt (for example in the British Milk Marketing Scheme). In addition, the completion of the single market will have important consequences for U.K. agrifood manufacturing firms, and for the U.K. food consumer.

STRUCTURAL ASPECTS OF THE U.K. AGRIFOOD MARKETING SYSTEM

(a) Structural Concentration and Changes in Distribution in the U.K. Food Marketing Sector

Much academic interest has recently been focused on two important aspects of the structure of the U.K. food marketing sector. First, the types and extent of structural concentration occurring in both the manufacturing and retailing industries has generated much interest; *Burns, J. A. (1983); Carter, D. (1989); Duke, R. C. (1989); McDonald, J. R. S. et al. (1989);* and *Shaw et al. (1989).* Second, major changes in the way in which food is distributed to consumers has also become topical; *Burdus, A. (1988); Tanburn, J. (1981); Hunt, I. (1983),* and *Atkinson, G. (1986).*

Burns, J. A. (1983) traces the structural development of the U.K. food manufacturing sector from the early 1970s through to the early 1980s, and illustrates that over this period the sector has become increasingly dominated by larger firms. Of 5,000 food manufacturing firms in the U.K. in 1983, over one third of all U.K. food sales was already accounted for by just 10 companies. *Ashby, (1978)* and *Mordue (1983)* further demonstrate that from the mid 1970s to the mid 1980s the average size of U.K. food manufacturing establishments had become much larger than for manufacturing as a whole. *Burns, J. A. (1983)* also reports that the larger, dominant food manufacturers have responded to channel power concentration at the retail end of the food marketing chain in three ways:

- by rationalisation, as explained previously;
- by diversification, to lower dependence on specific markets; and
- by further spreading their interests outside the U.K.

The more recent history of the U.K. food manufacturing sector has been characterised by a slowing of this structural change process; *McDonald, J. R. S. et al. (1989).*

If the recent history of the U.K. food manufacturing industry has been a slowing down of structural change, the food retailing sector, by way of contrast, has undergone an acceleration in structural concentration. It was not until the 1950s that self-service food stores began to appear in the U.K. Since that time retailing in general, and food retailing in particular, has not merely changed; as *Hunt, I. (1983)* puts it, "it has been transformed." One of the most intense elements of this transformation, over the past decade in particular, has been the rise to prominence of a very few large food retailing multiples. In recent years, thousands of small food shops and regional chains have closed in the U.K., beaten on price and quality by the larger scale-efficient multiples. *Duke, R. C. (1989)* reports comprehensively on the decline of the smaller U.K food retailers and the rise of the "big five" multiples. As illustrated in Table 2, the "big-five" now account for over 50% of the total U.K food retail market.

From Table 2 it can be seen that the multiples as a group control 78% of all U.K. food retail sales. This represents a considerable concentration of retailing activity, and it gives the multiples enormous bargaining power over the food suppliers. This power, often described by U.K. food manufacturers as "retailer tyranny," is such that suppliers can be compelled, for example, to cut prices, supply own-label goods, and pay for in-store promotional activity. What has been witnessed in the U.K. food marketing sector over the past decade has been an accelerating process of oligopolisation of the food retail market. This means that, while a large multiple retailer deals with a large number of suppliers and potential suppliers, each responsible for only a small proportion of its total turnover, the suppliers are increasingly faced with a relatively small number of significant buyers, the loss of any one of which is likely to represent a major loss of trade. Whilst this process of retail oligopolisation obviously benefits the multiples, and arguably also the food consumer in the form of price control, it does however represent a major threat to individual food manufacturing firms.

Duke, R. C. (1989) reports on how U.K. food manufacturers are spending increasingly heavily on promotion in an effort to build brand loyalty. Such investment in brand loyalty is not just aimed at

Table 2: U.K. Food Retail Market Shares: 1989/90

Company	% Share	
*Sainsbury	13.7 }	
*Tesco	12.9 }	
*Gateway (Dee)	11.3 }	54.8
*Argyll	9.5 }	
*Asda	7.4 }	
Other Multiples	23.2	
Co-ops	12.9	
Symbols	3.6	
Independents	5.5	
	100	

* The "Big-Five" Multiples

Source: Audits of Great Britain/Mintel

competing with rival manufacturers; rather it can be seen as an effort to counteract food outlet loyalty (now the domain of the multiples), and thus ameliorate the present imbalance in monopsonistic channel power existing in favor of the multiples. Thus, the U.K. food shopper has entered into a very dynamic and exciting retail battlefield in which the multiples and the food manufacturers fight for power over each other.

(b) The Role and Future of the U.K. Agricultural Commodity Marketing Boards

One area in which the U.K. agrifood marketing system is certainly unique (and many would say anachronistically so), amongst its E.C. partners, is in the important role still played by some Commodity Marketing Boards. Opponents of the whole concept of Marketing Boards, and the interventionist Agricultural Marketing Acts which enabled their establishment, include *Pool and Threipland (1989)* who state that "since 1973, when the U.K. joined the E.C., national producer boards have become anachronisms." Debate at present revolves around the compatibility of these virtually monopolistic bodies with the tenets of E.C. competitive philosophy. With 1992 almost upon us, this debate has now reached fever-pitch in some sectors, such as the British Milk Marketing Scheme.

In the context of this descriptive article, it will be useful to explain briefly the background to the establishment of the various boards, to outline their statutory functions, and finally to investigate some of the more compelling arguments, both in favor of, and contrary to, their continued existence.

Like all policy instruments, the Agricultural Marketing Acts of 1931 and 1933, which allowed for the establishment of the commodity boards, can be seen as having a parentage of practical pragmatism and political expediency. The precise balance of pragmatic and expedient elements is a topic for debate outwith the scope of this paper. On the practical side, the Acts, and the boards which they produced, can be viewed as well-intentioned efforts by Ramsay McDonald's Labor government to ease the hardship of farmers in the depressed agricultural sector of British industry during the interwar years. In terms of expediency, it can be argued that McDonald's government, in 1931, desperately needed all the support it could get in rural by-elections As a result, McDonald's cabinet "gave in" to (a then fairly powerful) farm lobby, and passed the 1931 Act, to be followed by the same procedure two years later. Whatever one's interpretation of the pragmatism or expediency of McDonald's government in passing the 1931 and 1933 Acts, they certainly represented the beginning of a new era of government intervention, control and assistance to U.K. agriculture. As a direct result of the Acts, compulsory commodity boards were established for hops (1933), milk (1933), pigs and bacon (1933), potatoes (1933), and milk produce (1939).

There is still considerable dispute over whether farm gate prices were in fact rescued by the marketing boards, or whether the subsequent price improvements were the result of general economic recovery. Certainly, the boards brought the interwar farmer the highly valued benefits of a guaranteed sale at the same price as his neighbour together with prompt, assured payment. The Centre for Policy Studies indicates that in the milk sector, for example, U.K. milk producers, served by their milk marketing boards, receive only 86% of the average E.C. price; *The Economist, (10/3/1990)*. However, complex issues arise when attempting to make a comparison of this kind which are discussed in Ritson and Swinbank (1991). Yet prices to U.K. consumers also tend to be higher than in the rest of the E.C.

The marketing boards still in existence are for:

- milk (England and Wales, one board; Northern Ireland, one board; Scotland has three boards);
- wool (a single Board for the whole U.K); and
- potatoes (for ware potatoes in Great Britain only. Northern Ireland has no Board for ware potatoes, although it did have a seed potato marketing board during a brief period in the 1970s).

Of these several boards, the Milk Marketing Board for England and Wales is by far the most important. In its function as sole purchaser of milk, it has a current turnover of £2,000, million, and with its commercial activities, through Dairy Crest Ltd, a further £750 million. To put this in context, this Board's total turnover is equivalent to 9% of the total U.K. expenditure on food in 1990.

The "monopoly" of the milk boards rests upon the fact that U.K.

milk producers are prevented, in law, from selling any of their milk to anyone except the local milk board (with certain exceptions due to E.C. legislation). Equally, the milk marketing boards are required to buy any milk that dairy farmers within their regional jurisdiction legally produce, having due regard for E.C. milk quotas. The Dairy Trade Federation, representing the private milk processors, negotiates directly with the milk marketing boards for its raw milk requirements. However, an anomaly exists here, in that the English and Welsh MMB owns a processing firm, Dairy Crest Limited, which currently gets over one-third of all raw milk supplies. This anomaly has raised questions of fair competitive practice.

Another well documented aspect of the current system exists in the way in which liquid milk gets priority over other uses when supplies are short; *The Economist, (10/3/1990); McOueen, J. D. W. (1990); Wilson, C. H. (1991); The Economist, (13/10/1990); Groves, C. R. (1982); Agra Europe, (1989)* and *Meynell, P. J. (1990).* Next in priority comes specialty products, then cheese, then butter. The result of this arrangement is that in late summer, when cows are dry, supplies for processing are restricted and processing plants can lie idle. This problem has been exacerbated by the reduction in overall U.K. milk production following the imposition of quotas. The net impact on the U.K. food market is that increasing quantities of milk-based, value-added products are being drawn in from France, Italy, Germany, Switzerland and Denmark.

However the most contentious issue surrounds what is known as "end-use" pricing in which the price of milk allocated to different uses is calculated according to a formulae which deducts marketing and processing costs from the estimated end value. The "natural monopoly" provided by a ban on imports of liquid milk (for "health and hygiene" reasons) together with a very inelastic demand, allowed the price of milk to the liquid market to be much higher than for milk used to manufacture milk products. (Farmers receive a "pool price.") This has become known as the "liquid premium"; it was expected to disappear (or at least contract) following the European Council rulings which have removed the import ban; the Board has remained relatively successful in sustaining the liquid premium from the liquid market. (See Table 3.)

The past two or three years have seen more speculation and

Table 3

Milk Prices (pence per litre)

	April 1991	% change compared with 1989 peak
MMB Selling Prices:		
Liquid	24.745	+11.0*
Butter		
SMP aided	15.119	-16.1
SMP unaided	18.375	-3.5
Cheese		
Cheddar	15.875	-14.0**
Short life	15.875	-15.4
Condensed milk	18.600	-3.9
Evaporated milk	18.600	-1.3
Chocolate crumb	18.800	-3.7
Whole milk powder	17.650	-8.8
Heavy cream	18.200	-4.9
Target Price	20.93	
Producer price	18.776***	
Producer-Retailer Contribution	4.531	
Producer-Processor Contribution	4.945	

* compared with approximate average for the year
** subject to arbitration
*** average price paid March 1990 to February 1991

Ritson, C. and Swinbank, A. (1991)

uncertainty over the future of milk marketing in the U.K. than at any time since the period directly before the U.K. joined the European Community in 1973, when there was serious doubt over whether the Milk Marketing Boards could continue in existence. Recent anxiety has been caused, at least in part, by concern over whether the British Milk Marketing Scheme can truly be compatible with the "1992" programme. *Ritson and Swinbank, (1991)* provide a comprehensive discussion of the basic conflict between the British Milk Marketing System and the competitive philosophy of a common market, and explore how the U.K. milk marketing system might evolve as a consequence of completion of the internal market. The central issue in relation to completion of the single market relates to the nature of competition policy within the U.K. system. Can the E.C. authorities really be satisfied that private, British based, milk product manufacturing companies are able to compete effectively in any single market, given that the English and Welsh Milk Marketing Board owns a major processor, Dairy Crest Limited, and the MMB is the sole supplier of milk?

This issue has fuelled a major debate between the Milk Marketing Boards (MMBs) and the Dairy Trade Federation (DTF). The DTF argues that in relation to E.C. competitive philosophy, the possession of Dairy Crest Limited by the Milk Marketing Boards represents a wholly unfair situation. The British National Consumer Council has suggested that the Agricultural Marketing Acts should be repealed by 31st December 1992, effectively ending the Milk Marketing Board's monopolistic position in the U.K. Milk Marketing Scheme, and that Dairy Crest Limited should be floated on the stock market as a private manufacturing company, putting it on the same footing as other members of the DTF.

At the time of writing (January 1992) the MMB has just announced its intention to convert itself into a voluntary Co-operative with Dairy Crest becoming a separate company, owned by producer shareholders. This change was virtually forced on the MMB by discrete Government pressure together with a realization that the European Commission was willing to "strengthen" the scheme to prevent low fat milk by-passing the MMB.

The British Wool Marketing Board was formed in 1950 as part of the post-war reconstruction of Britain. It has a monopoly of the purchases of fleeces of sheep from the farmers whose animals produce them. There exists a *de minimus* exemption for the sale of a small number of fleeces. The Board currently has a turnover of less than £60 million. It employs merchants, often part or wholly owned by itself, to grade and sort the wool; and it then sells the wool by auction on the world market. Since the foundation of the Board, the British government has supported producer wool prices, initially by subsidy; at present by a guarantee system which effectively smooths price fluctuations year on year. Although this guarantee will stay for perhaps another couple of years, the present government is phasing it out. Thus, the industry will be without government subsidies, and will rely on the world market to determine wool prices directly. However, there appear to be no moves by the government to end the monopoly position of the Board in buying producers' fleeces.

The Potato Marketing Board operates only in Great Britain, not the whole of the U.K. It operates a system of area quotas; producers have an allocated area or quota and pay a levy on that area. For plantings in excess of allocated quota, the levy is multiplied by a factor of ten. There are also arrangements for trading in potato quota. Unlike the Milk Board, the PMB is not involved directly in the marketing of the potato crop, but has operated a complex system of market support, buying surplus production either by pre-season contract with growers, or by spot purchasing, and selling the surplus cheaply for cattle feed. This support has traditionally been jointly financed by growers and government. However, in 1989 the government withdrew its funding of support to the potato market and undertook a major review of the Board. Surprisingly, the PMB survived with the introduction of industry and consumer representation, and growers and buyers required to contribute more support, funding a ban on imports. However, E.C. membership ended the total ban, and in 1979 prohibition on health grounds was successfully challenged by a Dutch potato trader, and there now exists totally free imports of potatoes to the U.K. Over the past few years, therefore, two of the three original pillars of the Great Britain potato marketing scheme have been removed–government price subsidy and import prohibition. What remains is the production quota system–but even this is under renewed threat, following attempts to introduce an E.C. potato regime under the 1992 programme.

Thus, the future form, if not the very existence, of the remaining British commodity marketing boards is now a matter for serious consideration in U.K. agrifood marketing circles, for three reasons. First, in that the three boards restrict competition in one form or another, there is a question mark over their acceptability under the "1992" programme. Second, some producers are questioning whether their performance may truly benefit agricultural producers. And third, given that many of the original functions performed by the boards have disappeared, some people argue that there is no reason why such emasculated and anachronistic bodies should be maintained.

(c) The Role and Status of Cooperatives in the U.K. Agrifood Marketing System

Although the co-operative agricultural marketing sector continues to grow in the U.K., albeit rather slowly, it is acknowledged by experts in the field that co-operation in the U.K. has never reached the same relative proportions, in terms of either membership, turnover or market share, as in other E.C. countries; *Bailey (1985), Baron (1970), Foxall (1982) and (1984)* and *Morley (1975)*. The most recent statistics provided by the "Plunkett Foundation for Co-operative Studies," show that there are currently a total of 643 registered agricultural co-operatives in the U.K. Year on year growth in numbers over the past decade has been static at about 1%; *Millns, J. 1990*. Millns reports that the seven new co-operatives established in the U.K. between 1989 and 1990 are marketing groups in the product areas of cereals, vegetables, fruit, peas and beans. Pessimistically, Millns forecasts that it is unlikely that many more marketing groups will be formed over the coming decade. One of the main reasons for general stagnation in the U.K. co-operative marketing sector, and presumably also accounting for Millns' pessimistic forecast, is that more producer members are finding traditional cooperative legislation restrictive. This is particularly true in respect of their ability to access capital for new developments. Existing U.K. marketing co-operatives are struggling to invest in much needed capital equipment to enable them to compete effectively in a single Europe. *Millns (1990)* reports wholesale financial and corporate restructuring of large numbers of U.K. agricultural co-operatives, in an attempt to access capital and remain, or become, competitive. Many groups have ceased trading as co-operative societies and have reconstituted as private limited companies. Yet others have restructured operations by decentralising activities and establishing independent limited companies, within the parent co-operative, as individual profit centres.

Under pressure from dissatisfied producer members, combined with the need to prepare competitively for "1992," and restricted financially by anachronistic legal structures, the U.K. agricultural co-operative sector is currently going through a major crisis. An expert on co-operative legislation, *Ian Snaith (1988)* lays the blame for the crisis at the door of the organisations responsible for formulating and revising co-operative legislation. In terms of legislative development, Snaith characterizes the 1980s as "a decade of neglect."

Another, perhaps gloomy, development in the co-operative sector has been the radically altered remit of Food From Britain, the umbrella organisation for all U.K. agrifood co-operatives, and also responsible for the generic promotion of U.K. agrifood products in export markets. In 1990, Food From Britain distanced itself from agricultural co-operation for the first time since its inception, by stating that "efforts would not be focused so much on co-operatives, per se, but on the creation of any effective structures for marketing agricultural produce, particularly in export markets"; *Millns (1990).* This represents another aggravating element in the crisis facing the co-operative sector. Total sales turnover of the sector currently stands at £2.8 billion (less than 15% of the total spent on U.K. agrifood products by U.K. consumers in 1990).

Assuming that U.K. agricultural marketing co-operatives are achieving their general aim of improving the marketing of farmer members' agrifood products, then it is clear that if they are to continue in existence, rather than be converted into private limited companies or collapse through an inability to compete in Europe, they must prevent the 1990s from becoming another "period of (legislative) neglect."

CONCLUSION

At the time of writing, many changes, some radical and some not so radical, are occurring within the U.K. agrifood marketing system. This article has attempted to provide an eclectic, and yet topical, discussion of some of these major changes.

Within the *marketing environment* facing the U.K. system, a "consumption revolution" is taking place in U.K. households, which has far-reaching consequences for farmers, food manufacturers and retailers alike in adapting to provide for new consumer tastes.

A second major aspect of the *marketing environment* which is ushering in changes to traditional thinking and practice, is the impending single European market. "1992" will have a profound effect upon the structure and practices of specific sectors, such as milk and livestock, whilst at the same time posing potential opportunities and threats for both U.K. food manufacturers and food consumers.

The *structure* of the U.K. agrifood marketing system is also undergoing transformation. In particular, we are witnessing extreme oligopolisation of the food retailing sector, and the continuation of a major brand loyalty battle between manufacturers and the multiples. The future roles and structures, and indeed the continued existence of, the three remaining commodity marketing boards are also under active consideration at the present time. And finally, the agricultural co-operative sector, never as strong or as large in the U.K. as in other European countries, has woken up to find that it has shot itself in the foot. Through this self-inflicted wound, born of neglect, co-operatives are slipping out of the co-operative fold and into the more dynamic world of private limited companies.

REFERENCES

Agra Europe. (1989). *Pricing Obstacle to U.K. Dairy Product Innovation*. Agra Europe, Nov. 17, 1989, Nl-N3.

Ashby, A. W. (1978). *Britain's Food Manufacturing Industry and its Present Economic Development*. Journal of Agricultural Economics, 29, 213-224.

Atkinson, A. G. (1986). *Competition in the Food Industries*. In J. Burns and A. Swinbank (Eds.) *Food Policy Issues and the Food Industries*. (pp. 63-87). Reading: University of Reading.

Bailey, D. G. (1985). *Agricultural Marketing and Cooperation: A Decade of Progress*. Farm Management, 5, 471-479.

Barker, J. (1989). *Agricultural Marketing. 2nd Edition*. Oxford: Oxford University Press.

Baron, P. J. (1970). *Why Cooperation in Agricultural Marketing?* Journal of Agricultural Economics, 29.

Beardsworth, A. (1990). *Trans-science and Moral Panics: Understanding Food Scares*. British Food Journal, 92(5), 11-16.

Boakes, N. (1990). *Trading with our European Partners*. Food Policy, 15(2), 161-166.

Burdus, A. (1988). *Competition in the Food Distribution Sector*. In J. Burns and A. Swinbank (Eds) *Competition Policy in the Food Industries*. (pp. 68-102). Reading: University of Reading.

Burns, J. A. (1985). *A Review of Food Policy Issues in the U.K.* Food Marketing, 1(3), 3-20.

Burns, J. A. (1983). *The U.K. Food Chain with Particular Reference to the Inter-Relations Between Manufacturers and Distributors.* Journal of Agricultural Economics, 34(3), 361-378.

Carter, D. (1989). *Developments in Grocery Retailing.* British Food Journal, 9(1), 13-15.

Cohen, S. (1970). *Folk Devils and Moral Policies.* London, Paladin Books.

Commission of the European Communities (1988). *Completing the Internal Market: An Area Without Internal Frontiers.* The Progress Report Required By Article 8B of the Reaty, COM (88), 650, CEC. Brussels.

Dancey, R. J. and Jones, G. L. (1990). *The Single European Market: "1992."* Farm Management, 7(5), 225-236.

Duke, R. C. (1989). *A Structural Analysis of the U.K. Grocery Retail Market.* British Food Journal, 91(5), 17-22.

Evanson, M. (1990). *Implications of 1992 for the U.K. Dairy Sector.* Food Policy, 15(2), 132-144.

Foxall, G. R. (1984). *Cooperative Marketing in European Agriculture: Organisational Structure and Market Performance.* International Marketing Review, Spring/Summer 1984, 42-57.

Foxall, G. R. (1982). *Cooperative Marketing in European Agriculture.* Aldershot: Gower Publishing Company.

Galama, K. (1990). *Meeting the Challenges for the U.K. Food and Drink Industry in a Single European Market.* BNF Nutrition Bulletin, 15, 71-75.

Gardner, B. (1989). *E.C. Dairy Policy to 1992.* Journal of the Society of Dairy Technology, 42(4), 106-111.

Gofton, L. R. (1990). *Food Fears and Time Famines: Some Social Aspects of Choosing and Using Food.* In Why We Eat What We Eat, Proceedings of British Nutritional Foundation Annual Conference, London.

Gofton, L. R. and Marshall, D. (1989). *Unpublished Report to the Ministry of Agriculture Fisheries and Food.*

Gofton, L. R. and Hess, M. R. (1991). *Twin Trends: Health and Convenience in Food Change or Who Killed the Lazy Housewife?* British Food Journal, 93(7), 17-23.

Groves, C. R. (1982). *The Marketing of Milk and Milk Products in the United Kingdom. Marketing Report No. 1,* The West of Scotland College of Agriculture.

Hunt, I. (1983). *Developments in Food Distribution.* In J. Burns, J. McInerney and A. Swinbank (Editors) *The Food Industry: Economics and Policies* (pp. 127-141). London: Heinemann.

Jamieson, M. R. (1988). *Branding, Market Segmentation and the Consumer.* Issues in Food Policy, 1, 45-55.

Kerr, K., Dealler, S. F. and Lacey, R. W. (1988). *Materno-foetal Listeriosis from Cook-chill and Refrigerated Food.* Lancet, Vol. 2, 11-33.

Lacey, R. (1989). *Listeriosis*. In Goldring, O. (Editor), Salmonella and Listeriosis, EAG Report, London, (pp. 211-30).

Lacey, R. and Dealler, S. F. (1990). *Food Iradiation: Unsatisfactory Preservative.* British Food Journal, 92(1), 15-17.

Mackenzie, D. (1990). *The Green Consumer.* Food Policy, 15(6), 461-466.

McDonald, J. R. S., Rayner, A. J. and Bates, J. M. (1989). *Market Power in the Food Industry: A Note.* Journal of Agricultural Economics, May/June 1989, 101-108.

McQueen, J. D. W. (1990). *Milk Utilization and the Role of the Milk Marketing Board.* Journal of the Society of Dairy Technology, 43(1), 1-14.

Meynell, P. J. (1990). *Milk Producer Organisations in the United Kingdom: Structure and Member Relations.* Oxford: The Plunkett Foundation For Cooperative Studies.

Millns, J. (1990). *Agricultural Cooperation in the United Kingdom.* In Yearbook of Cooperative Enterprises 1990 (pp. 147-151). Oxford: The Plunkett Foundation For Cooperative Studies.

Mordue, R. E. (1983). *The Food Sector in the Context of the U.K. Economy.* In J. A. Burns, J. P. McInerney and A. Swinbank (Editors), *The Food Industry: Economics and Policy* (pp. 18-50). London: Heinemann.

Morley, J. (1975). *British Agricultural Cooperatives.* London: Hutchinson Benham.

Palmer, C. M. (1990). *The Single European Market and the Meat and Livestock Industry.* Journal of the Agricultural Society, University College of Wales, 70, 222-238.

Paulson-Box, E. and Williamson, P. (1990). *The Development of the Ethnic Food Market in the U.K.* British Food Journal, 92(2), 10-15.

Pickles, L. (1990). *1992 from the Perspective of the U.K. Food Manufacturer.* BNF Nutrition Bulletin, 15, 50-61.

Pool, R. and Threipland, A. (1989). *Set Food Markets Free: Repeal the Agricultural Marketing Acts.* London: Centre for Policy Studies.

Pooley, P. (1989). *Address to the European Dairy Trade Association (ASSILEC) Assembly*, Edinburgh, U.K. 16th June 1989.

Ritson, C. and Hutchins, R. (1990). *The Consumption Revolution.* Unpublished paper delivered at a Symposium celebrating 50 years of the National Food Survey. A shortened version appears. In "Fifty Years of the National Food Survey," ed Slater, J. M., MAFF, London, ISBN 011 242909 2, pp. 33-46 (1991).

Ritson, C. and Swinbank, A. (1991). *The British Milk Marketing Scheme: Implications of 1992.* 25th European Association of Agricultural Economics Seminar, Braunschweig-Volkenrode, Germany, June 1991, pp. 165-180.

Shaw, S. A., Burt, S. L. and Dawson, J. A. (1989). *Structural Change in the European Food Chain.* In B. Traill (Editor) *Prospects for the European Food System* (pp. 3-34). London: Elsevier Applied Science.

Snaith, I. (1988). U.K. *Cooperative Legislation in the 1980s: A Decade of Ne-*

glect. In Yearbook of Cooperative Enterprise (pp. 157-168). Oxford: The Plunkett Foundation for Cooperative Studies.

Swinbank, A. (1990). *The Economics of 1992: Food.* BNF Nutrition Bulletin, 15, 40-49.

Swinbank, A. (1983). *The Food Industry and the E.C.* In J. Burns, J. McInerney and A. Swinbank (Editors) *The Food Industry: Economics and Policies* (pp. 230-239). London: Heinemann.

Tanburn, J. (1981). *Food Distribution: Its Impact on Marketing in the 1980s.* London: Central Council for Agricultural and Horticultural Cooperation.

The Economist (13th October, 1990). *Bottling Out: The Dairy Industry.* p. 35.

The Economist (10th March, 1990). *Milk Shake-Up: The Dairy Industry.* p. 35.

Theroux, P. (1989). *Riding the Iron Rooster: By Train Through China.* New York: Ivy Books.

Thompson, J. (1988). *Farm-Based Food Processing Enterprises.* Issues in Food Policy, 1, 30-44.

Wheelock, J. V. (1990). *The Consumer and Agriculture.* Journal of the Agricultural Society, University College of Wales, 70, 138-150.

Wheelock, J .V. (1986). *Coping with Change in the Food Business.* Food Marketing 2(3), 20-45.

Wilkinson, G. A. (1984). *The U.K. Dairy Industry and the E.E.C.* In A. Swinbank and J. Burns (Editors) *The E.E.C. and the Food Industries* (pp. 86-95). Reading: University of Reading.

Wilson, C. H. (1991). *Developing and Marketing Dairy Product Brands in Europe.* British Food Journal, 93(4), 3-11.

Wilson. J. (1991). *Food for Thought.* Marketing Business, December/January 1991/92, 40-43.

Wright, G. (1988). *The Consumer Reaction to Food and Health Issues.* Issues in Food Policy, 1, 75-84.

Agricultural Marketing in France

Bernard Yon
Sylvie Bernaud

SUMMARY. French production is based on a large agricultural area with a large number of farmers. They work with co-operatives which are very near to them and collect their product. Protectionism has led to a disequilibrium with the economic system in this sector. Actual CAP's orientations and the increase in competition, has forced farmers and co-operatives to concentrate and restructure. The number of farmers has decreased while many co-operatives have started to process raw production materials.

One goal of the food industry is to become more international. French food industry made a lot of acquisitions abroad but the increasing place of Italians in French companies also has to be noticed.

Distribution plays an important role in agricultural marketing where 66% of food products are sold through about ten companies. This market structure allows them to have a favored position. However, it's a very dynamic sector which now faces some difficulties.

France is a country where agriculture represents an important part of the national economy. It is necessary to study agricultural marketing in order to better understand the interactions between the different actors taking part in the food chain from suppliers to consumers.

Bernard Yon is Professor at the Institut de Gestion Internationale Agro-Alimentaire, Centre Polytechnique Saint-Louis, Cergy-Pontoise. Sylvie Bernaud is Associate Professor at the Institut de Gestion Internationale Agro-Alimentaire, Centre Polytechnique Saint-Louis, Cergy-Pontoise.

[Haworth co-indexing entry note]: "Agricultural Marketing in France." Yon, Bernard, and Sylvie Bernaud. Co-published simultaneously in the *Journal of International Food & Agribusiness Marketing* (The Haworth Press, Inc.) Vol. 5, No. 3/4, 1993, pp. 113-125; and: *Food and Agribusiness Marketing in Europe* (ed: Matthew Meulenberg), The Haworth Press, Inc., 1993, pp. 113-125. Multiple copies of this article/chapter may be purchased from The Haworth Document Delivery Center [1-800-3-HAWORTH; 9:00 a.m. - 5:00 p.m. (EST)].

This article will present characteristics of the agriculture sector, cooperative system, food industries and distribution in France.

FRANCE: A COUNTRY OF SPACE

France is a country which has a large agricultural area. The potential agricultural area represents 64% of total French territory (about 31 millions hectares, it is the highest of European countries). Agriculture is still generally a function of space; that is not the case in other countries where agriculture is more intensive as for example Holland. French production is intensive but necessitates relatively less investment than an intensive industrial agriculture.

Yields obtained in France are some of the highest in the E.E.C. So France has a high potential mean of production.

A good comprehension of problems in this sector has to include the past. The agricultural sector is characterized by a disequilibrium with the economic system. Indeed, protectionism in the form of support prices has made farmers dependent on government regulation and control. Explaining problems of this sector by overproduction is only the reflection of this way of analyzing the situation. Indeed, Europe represents a market of 380 million consumers and French production is far from saturating the market. This way of thinking based on self-sufficiency of the French market makes no sense because the European economy's space is now homogeneous. The key explaining the situation is, as for every good, the laws of the market. French agricultural production is in competition with that of other European countries and it is this environment of high competition that agriculture has to face.

The CAP's orientation is as follows. One of the five major objectives (article 39 and 40 of the treaty of Rome) was to obtain self-sufficiency in food for Europe. This was reached 20 years ago. The actual trend is the reduction of costs created by this policy. The CAP is now turning towards the reduction of price supports, leaving market laws to act more normally. However the policy of cost reduction of the CAP is different depending on the sector. It imposes reduction of quantities (quotas) for milk, for instance, or reduction of prices for cereals. These two orientations have different impacts on the sector concerned. Indeed, reduction of price leads to a quicker and safer result because it allows output restructuring based on a competitive market. The policy of quotas allows little farms to survive (i.e., an important social factor), has less dynamic effects regarding restructuring, and leads, in the long term, to the diminution of the capacity for production. This consequence can be harmful as recently seen. The selling of 1700 million of dollars of cereals to the USSR decreased stocks to a level never before reached in 15 years. Thus the potential for greater production has to be kept open with the possibility of reducing production when necessary.

The evolution of the agricultural economy is characterized by 2 phenomena. First, there is a constant and rapid expansion of general agricultural production and exchange. The active population in agriculture decreases regularly and this reduction is accelerating at the same time as the volume of production increases. This fact leads to a quick increase in labor productivity. Active farmers currently represent 6% of the total workers population. The second phenomenon is that the value of agricultural activities has seen its share in the total value of the economy decreased. The added value of the agricultural sector doesn't increase as quickly as the raw gross national product. (It represents now 2.9% of the raw gross national product.)

Concerning business concentration, it is interesting to notice that it can lead to only weak economies of scale. Only some parts of the capital costs amortized over time will decrease but variable costs for inputs such as seed quantity, pesticides and fertilizer quantities as well as working hours or energy are proportional to the activity. So concentration alone doesn't create a higher profitability of the inputs (Table 1). The actual average surface of a farm in France is 36 ha.

In conclusion, a producing country has to produce at the lowest costs. This competitivity has to be at its maximum all along the marketing chain: from the farm to product collection and through marketing to its final outlet.

THE COLLECTION SYSTEM

In France, 3/4 of agricultural production is collected by co-operatives and one fourth by private firms (Table 2). The actual trend in

TABLE 1. Example of costs structure for wheat production in France, Marne, 1991.

	FF/ha	%
fertilizer	1170	11,96
seed	506	5,17
chemical	1213	12,40
labor cost	1953	19,96
mechanism	1867	19,08
construction	285	2,91
financial cost	1272	13,00
land rent	729	7,45
general cost	788	8,05
TOTAL	9783	100

TABLE 2. Average quantities of agricultural produce collected and evaluation of cooperative's structure, France 1970/1990.

	1970	1980	1989	1990
average per coop (in q)	231,500	432,500	730,000	1,500,000
Number of coop producing:				
less than 300 000 q	543	358	208	98
300 to 1 000 000 q	104	136	135	66
+ thán 1 000 000 q	24	56	94	69
total	671	550	437	233

q = 100 kilograms

this sector is for co-operatives to merge and if the tendency continues, 50% of currently existing co-operatives will have disappeared in 5 years. This concentration allows larger investment with an increase in profitability due to economies of scale.

French co-operatives are very much involved in the agricultural sector. They are mainly locally based and are very careful about the environment and with the preservation of rural life. Co-operatives are linked with farmers and have got a way of thinking nearer to that of farmers than to that of industry. Because of this characteristic, co-operatives have a "follower" rather than a leader strategy. As they didn't invest and restructure earlier, they now have to face continuously decreasing profitability. At the present time, their lack of assets doesn't allow them to invest as quickly as the economic evolution would necessitate.

The co-operative reform of 3 January 1991 tried to ameliorate this situation with the objective of bringing the co-operative's structure closer to that of a public enterprise. This reform can be summarized in 4 main points:

- It will make linkages with the private sector and the creation of branches more easy. Capitalistic (private) investment into agricultural companies of collective interest (SICA Societé d'Intéret Collectif Agricole) is currently very difficult. Every SICA which has been originally created to favor partnership between farmers and a co-operative on the one hand, or a co-operative and industry on the other hand, will now have the possibility of being transformed into a limited company.
- This reform will allow the search for new financial assets. The possibility of issuing debentures, which has been suppressed since 1967, was re-established in order to facilitate co-operative access to the financial markets. Initial capital investment in cooperatives can also be reinforced by the issue of co-operative certificates of investment.
- Rules of investment participation and interest benefits for employees can also be applied. Co-operatives are allowed to distribute dividends to their members in different branches.
- New legal and tax constraints will probably lead to a progressive abandonment of the SICA statute.

THE ROLE OF CO-OPERATIVES IN PRODUCT TRANSFORMATION

The restructuring of co-operatives expresses the need for an organization based on new principles and new organization.

Co-operatives have an economic goal and want to go farther in the transformation of agriculture products and in international exchange. They now need to develop at the level of the new Europe. The time when members were close to their co-operative is now changing. Actual cooperative strategy aims to diversify and penetrate the agri-industrial sector. Indeed, 80% of agricultural production goes to industry. So co-operatives have to develop their own transformation activities in order to increase their share of this sector and increase their margins. We have to point out that if local agricultural products are dependent on industry for their commercialization, the opposite is not true. In case of disruption in supply in the French market, industry can easily supply their needs from the world market. Co-operatives however, remain dependent on agri-industry.

Co-operative strategy of increasing their role in transformation has not only the goal of increasing profitability but also of creating a better unity all along the French food chain. Let us introduce some examples illustrating this search for co-operative diversification:

- The UNCAC (National Union of Cereals Co-operatives) has created a research and financial company (assets of 500 millions FF). The study of market diversification for cereals co-operatives is the main goal of this company. It is also exploring the possibilities of participation in food industries.
- The Pau co-operative has turned its efforts to vegetable production and transformation. Under the trade name "Green Giant," 80 million cans are produced; and under the trade name "Bonduelle" (a 50% partnership) canned and deep-frozen food products are turned out.
- The Cher Union count on new products and developed 3 lines of diversification: millery, bakery and animal nutrition.

This strategy presents risks. As they become more involved in transformation, co-operatives will have to pursue a strategy based on the key role of costs. The laws on distribution are very strict.

They only allow about a 2% price increase per year. But sale bargaining very often ends up in 1.5% added rebate, leaving only a 0.5% real price increase to the producer. Taking into account a 2% inflation rate, food industry has to carry out productivity gains of about 1.5% each year if it wants to maintain its profit margin.

In this way, technological know how and investment are both indispensable parameters for success in this sector. By attempting to penetrate the transformation market, co-operatives are in direct competition with the agri-industry. These industries are more aware and used to the rules of distribution and they represent a hard competitor.

THE FOOD INDUSTRY

France is a country where food tradition is very important and represents a national asset. This fact perhaps explains the development of French food companies. They are still not very developed internationally abroad. However, in 1990, firms tended to re-orientate their activities to find partnerships in other countries. We can point out that French companies have invested more money abroad than foreign companies have done in France (12 500 million FF against 6 000). And the major part of this investment has been made in the EEC (three fourths). In 1990, French food industries have shown more modesty than the previous year. Transactions consisted of less expensive acquisitions than in 1989. This shows the actual trend for internal restructuring on basic product markets.

Partnership has also increased. It allows less expensive investment, mostly in non-capital investment; it reduces individual financial contributions and it allows the pooling of complementary skills.

Among the main investments carried out by French companies abroad in 1990, we can point out the following examples:

- BSN bought Birkel, a German company with 835 million FF assets and Agnesi, an Italian company of 500 million FF assets.
- Frealim (part of the Saint Louis Company) bought Frudesa (900 million FF assets) which is the leader of pre-cooked dishes in Spain.

On the other hand, French food industry is more and more connected with Italian. Italian companies recently took an increasing part in this sector, buying or increasing their shares in food companies. The Agnelli's family (IFI company) has now a substantial share in BSN. After having about 6%, they now reached 25% of the total assets of BSN becoming the most important share-holder. The IFINT (another branch of IFI company) just acquired Exor Holding and Perrier.

Mister Raul Gardini (with the help of JM Vernes) acquired a substantial share of Sucden (Cacao Barry, Sogéviande). These two examples show the increasing role played by Italians in the French food industry.

The growth in exports is difficult to confirm. Even if the food balance is positive, exchanges are characterized by the export of raw and intermediate products whereas imports consist mainly of finished products.

Only specific products like champagne with a very strong trade name image, have an important export market. Indeed, Moët et Chandon export 80% of total sales of champagne. Few other goods are in the same situation.

The impact of the AOC (Appelation d'Origine Contrôlée: label certifying specific characteristics of a product) on European consumers may be the solution for the 4 000 French food industries to survive face to face with giants such as Nestlé or Unilever.

Intermediate products also represent a sector with good prospects. They make up a major part of the food industry. In fact, food is more and more elaborated and it demands the incorporation of an increasing number of ingredients such as jellifients, emulsifiers, flavors and preservatives.

Food production requires a complex and varied technology. Innovation also is important in the food industry and subcontracting (biscuit topping for instance) spreads because it permits a more flexible response for the finished food products industry. The Roquette company is an example of an intermediate product company which has seen a large development. It provides products such as starch and substitute sugar products to food industries. An advantage held by this firm is to avoid problems associated with product distribution. In fact, the food market which links industries to distribution is very selective and competitive. The distribution company has a lot of power and negotiations are very difficult.

THE DISTRIBUTION

Large shops represent a more and more compulsory outlet for food product industries. Whereas 73% of non food products are sold through small non-organized structures, only 1/3 of food products are marketed in this way.

For food products, 28.2% of sales are done by hypermarket and 29.7% by supermarket. With popular smaller shops and co-operatives, it brings to 66% the share of large concentrated distribution companies in the sale statistics of food businesses. This forces industry to have a policy of large volumes.

The French distribution sector remains a very defensive sector. Many hyper- and supermarkets face a financial uncomfortable position which forced them to readjust their strategy. Indeed compared to the profitability of English distributors, the French profitability is smaller (1.5% for Carrefour against 7% for Marks and Spencer). This difference can be explained by French distribution policy: distribution companies favor large volumes and small margins while the English have an opposite strategy. The second explanation for this defensive position comes from an under-capitalization problem. Ownership capital of Carrefour equals 10% of its sales while suppliers' debt represents 170% of its personal assets. This situation is not very safe and can be very dangerous in case of changes in business cycle. The actual danger is the arrival of the German hard discounting firms (Aldi, Lidl).

Big chain distributors now have to face 3 challenges: First of all, competition is now more important because of the diminution of new location sites. The answer to this problem is given by a new way to publicize oneself. Monoprix chose a series of prices while Casino chose discounting.

The second challenge is the incidence of higher concentration in this sector. Whereas there are only about 10 firms, restructuring is very quick (Casino bought La Ruche Meridionale, Promodès bought a part of Codec and Carrefour bought Euromarché). The general manager of Promodès estimated that in the future only distributors with a volume of sales around 100 000 million FF will survive.

The last challenge is that of internationalization. Two countries are favored by French distributors: Spain where Promodès, Auchan and Carrefour are firmly settled but also the United States where results are, however, disappointing.

Other answers are given to the modification of the competitive environment. First, distribution tries to integrate industry. Transformation represents a large part of Casino's activities; it exists, too, in Intermarché and in Leclerc for the meat sector.

The development of a distributor's trade name is another solution. Indeed, the distribution trade name represents about 15% of the sales. Furthermore, an interesting line is the creation of a super European buying center such as Carrefour Metro.

In conclusion, French distribution is characterized by the presence of 5 solid groups with a European vocation: Carrefour, Auchan, Leclerc, Promodès and Intermarché. But some other weak structures remain: Unico, Cora, Rallye, Paridoc, Casino.

Distribution is a highly competitive sector and the problem of saturation has become more important because of the Royer law which controls business town planning.

The behavior of the french food consumer is evolving (Table 3). There is still a majority of consumers following the traditional way of eating: Three meals (breakfast, lunch and dinner) with a very structured composition of the two main meals (starter, main dishes, cheese and dessert). But this group of people is decreasing (48% in 1988 against 51% in 1985). A group which doesn't respect any law in the way they eat is growing (17% in 1988 against 13% in 1985). This group of people very often miss a meal (81%), eat between meals (84%) or consume 'light products' (82%).

CONCLUSION

Agricultural marketing in France includes the study of three markets. The agricultural market which links farmers to industry by the intermediate of co-operatives, the food market between industries and distributors and finally, the retail market between distributors and consumers (Figure 1).

TABLE 3. The French Food Consumer.

	1979 Expenditure		1989 Expenditure	
	% of total	% of food	% of total	% of food
food expenditure	24.5	100	20.3	100
at home	21.4	87.3	16.8	82.8
out	3.1	12.7	3.5	17.2

source: INSEE

The agricultural market is international. Agricultural products are sold and bought in global trading. The industrial sector is characterized by the important development of intermediate products.

The food market is national or deals with Europe. The European market is, as we saw above, the actual goal of French companies.

At least, the consumers' market is very local. So, one important characteristic of agricultural marketing in France, as elsewhere, comes from the taste of French consumers. Even if production technologies used are identical to those of other countries, consumers are different. Food industries are aware of this fact. Indeed, they don't use the same flavor for yogurt depending on the country of destination. The variety of beans for canning is different too.

In conclusion, we can say that agricultural marketing is characterized by a very specific structure: numerous farmers and consumers but only about 4000 food industries and 5 important distributors. This structure is the key to understanding the different relationships and forces existing between the different actors.

Figure 1: The French agricultural marketing

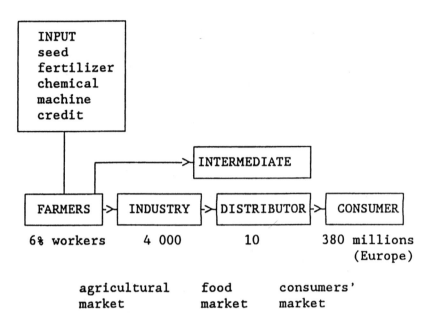

```
INPUT
seed
fertilizer
chemical
machine
credit
```

INTERMEDIATE

FARMERS -> INDUSTRY -> DISTRIBUTOR -> CONSUMER

6% workers   4 000   10   380 millions
                                               (Europe)

    agricultural   food   consumers'
    market   market   market

production <-> processing <-> selling <-> consumption
collection

REFERENCES

Cassignol, F. (1991). Cooperatives et négoces: fusion à grande vitesse, Cultivar, May, 296, pp. 25.

Gotier, J. (1991). Les relations commerces industries, Revue Industrie alimentaire, March, 458, pp. 20-22.

Jicquel, J. L. (1991). Recentrage et partenariat ont fait recette en 1990, Revue Industrie alimentaire, January, 455, pp. 22.

Lahon, J. (1991). Le pôle alimentaire Casino, Revue Industrie alimentaire, September, 466, pp. 14.

Manon, P. (1991). Les AOC à la rescousse des industries agro-alimentaires, Revue Industrie alimentaire, June, 463, pp. 52.

Moulin, F. (1992). La tradition alimentaire, la tradition perd du terrain, Les marchés, May, 34, pp. 32.

Rouaud, P. O. (1991). La distribution en révolution, Revue Industrie alimentaire, September, 466, pp. 30.

Rouaud, P. O. (1991). Distribution: un secteur défensif, Revue Industrie alimentaire, February, 456, pp. 29.

Approche du prix de revient du blé, Club demeter, 31 mai 1991.

Agricultural Marketing in Ireland

Eamonn Pitts

SUMMARY. In this paper agricultural marketing in Ireland is described, including (1) a brief review of the marketing environment, (2) a description of the marketing structures, in particular the changing role of producer cooperatives, (3) details of marketing practices in relation to price, products, promotion and distribution, and (4) a subjective evaluation of how well the system works in achieving its objectives.

INTRODUCTION

The key aspects which mark IRISH Agricultural Marketing are:

1. its dependence on export markets for its main crops/products (over 85% for beef and over 70% for dairy products).
2. the extent to which its agriculture is dominated by grass based animal products, particularly cattle, (38.5% of agricultural output in 1990), milk (32.4%) and sheep (4.5%).
3. the seasonality of production from grass which determines product mix, and affects storage, distribution and the farmers' return.
4. a weakness in development of value added products, particu-

Eamonn Pitts, B Comm, MEcon Sc, is Head of Marketing at The National Food Centre, Dunsinea, Castleknock, Dublin 15.

[Haworth co-indexing entry note]: "Agricultural Marketing in Ireland." Pitts, Eamonn. Co-published simultaneously in the *Journal of International Food & Agribusiness Marketing* (The Haworth Press, Inc.) Vol. 5, No. 3/4, 1993, pp. 127-140; and: *Food and Agribusiness Marketing in Europe* (ed: Matthew Meulenberg), The Haworth Press, Inc., 1993, pp. 127-140. Multiple copies of this article/chapter may be purchased from The Haworth Document Delivery Center [1-800-3-HAWORTH; 9:00 a.m. - 5:00 p.m. (EST)].

larly in the two largest sectors, dairy and beef which are largely marked by commodity marketing and by a heavy dependence on intervention.

THE MARKETING ENVIRONMENT

Because of the dominance of exports relative to the home market,
the marketing environment of greatest concern to Irish agriculture
and food processing firms is that pertaining in export markets such
as the U.K and Germany. Nevertheless trends in the home market
are important for producers and for many food firms. EC policy in
relation to export refunds is particularly important for markets outside the EC.

Consumers

Trends in Consumer behavior and attitudes in Ireland tend to
follow those in the major European markets. For example it is only
within the last few years that there has been any significant interest
by Irish consumers in organic products. Other major European consumer trends which have made an impact in recent years, are the
demand for light and low fat products and for ready meals. Some of
the latest consumer trends in Ireland originate in the U.S. and U.K.,
rather than in Continental Europe. A shared language and media
(the viewership of British television is high and availability of
British newspapers is widespread) and considerable migration have
contributed to close ties between Ireland and the Anglo-American
world, which is reflected in quick transfer of some new food trends
in these markets. This is particularly true of the catering market,
where we have assimilated in a twenty year period franchised ham-
burger bars, fried chicken outlets and more recently pizza parlors.
The microwave revolution has also had a greater impact in Ireland
than in Continental Europe.

The Irish consumer shares many of the concerns in relation to
food reported in the Continental literature. O'Neill has reported
that, of eleven individual issues of possible concern about food,
antibiotics, hormones, artificial coloring and unspecified additives

and preservatives were the objects of greatest concern, while cholesterol, fat, sugar and salt were causes of lesser concern. "Artificial" ingredients were a much greater cause of concern than natural ingredients. (O'Neill 1988).

The Irish consumer is younger than his Continental neighbor and poorer. Twenty-eight and nine-tenths percent of the population in 1986 were under 15, compared with a Community average of 19.1%. The proportion over 65 years was 10.9%, compared with a Community average of 13.7%.

GDP per inhabitant in 1989 is estimated at 11,467 purchasing power standard units, compared with a Community average of 17,106, i.e., the Irish figure is 67% of EC average (EC Commission). The discrepancy in GNP per inhabitant (a more appropriate measure of consumer welfare) is wider, because the FP between GDP and GNP is wider than in other countries, due to the extent of transfers of profits abroad by multinational companies operating in Ireland.

As is normal with less affluent societies, the proportion of total consumer expenditure spent by households on food is greater than average. Table 1 shows the share of consumer expenditure in Ireland and in EC 12 in 1988 on foodstuffs and non alcoholic beverages.

TABLE 1. Share of Consumer Expenditure 1988 - Percentage

	Ireland	EC 12
Foodstuffs	22.7	17.6
Non alcoholic beverages	1.4	0.5

Source : EC Commission, 1991.

Foreign Competition

The Irish market is extremely open to foreign competition. Partly as a result of the small size of the country, and partly from the high degree of specialization of its agriculture, the range of foods offered by the native industry is more limited than in a large country like France. This pattern has accelerated in the decades since EC accession. With increased travel the range of food products demanded by consumers has grown, almost all of which are imported. Examples of mass consumption products which were almost unknown in Ireland twenty years ago are lasagna, fromage frais, broccoli, and kiwi fruit.

There is no overt preference for Irish food products except in the traditional areas such as dairy products, meat and alcoholic beverages. Here the strength of local companies and brands, combined with a consumer perception that the native product is good, has tended to keep the native market share extremely high.

Foreign produce is particularly strong in the processed food area, where brand advertising is important. The strength and advertising spend of multi-national firms such as Kellogg's makes competition from smaller Irish companies difficult.

Government Policies

As in other EC countries, the Government is constrained by its obligations under EC law from directly supporting the agricultural and food industries in a manner which would affect competition. The Common Agriculture Policy applies fully and there is therefore free trade. A state trading organization which had monopoly export powers in the years before EC membership has been turned into a dairy producers' cooperative and has lost its monopoly position.

The agricultural and food sectors are seen by the State as cornerstones of the economy. One of the objective of the National Development Plan 1989-1993 is "the development of firms based on natural resources including food and fish." State Expenditure places emphasis on upgrading of marketing, product development R & D and management expertise of firms with good development potential. There is considerable encouragement of these sectors in the belief that Ireland has a competitive advantage in food production. The Government policy objective is to convert these advantages into jobs in the Irish economy, where the unemployment level, partly for demographic reasons, is one of the highest in Europe.

The State supports through annual grants two specialist agencies, for promotion of Irish meat (CBF) and of fish (BIM). There is also a general export promotion agency (ABT). Each of these agencies helps Irish firms meet foreign customers, carries out research on foreign markets, attends trade fairs, conducts trade missions, etc.

The State also supports the development of horticulture in Ireland through a special development board, An Bord Glas.

STRUCTURE OF RETAILING, INDUSTRY AND TRADE

The retail grocery trade in Ireland is highly concentrated with two groups (Quinnsworth and Dunne's Stores) accounting respectively for 21.5% and 20% of the grocery trade. A co-operative chain of independent grocers, Super Valu, has a further 11.5%. Multiples have a 74% market share in Dublin, with Quinnsworth and its associate Crazy Prices alone accounting for 38.7% (Data from Attwood Research of Ireland).

The power and influence of these retailers on the market is therefore considerable and there are periodic allegations about the abuse of these powers in relation to credit terms, demands for merchandising, policies in relation to stocking of products, etc. Nevertheless, relations in general between the grocery trade and the food industry are positive.

The food retailing multiples in Ireland are highly professional and profitable. They are often a channel for new ideas for the food industry and are an example to it of good marketing and responsiveness to consumer needs. Firms which meet the requirements of the professional Irish retail multiple trade usually have little difficulty in dealing with those in other countries.

The individual firms in the Irish retail trade have different policies in relation to the extent of "own branding" of food. One of the major companies, Dunne's Stores, gives predominance to its own St Bernard brand while also stocking brand leaders. All of the other chains provide a wide choice among brands and include a limited number of "own brand" lines among these choices.

The dominance of two companies in the retail trade would be seen in other markets as a major problem for the food industry. It is not seen as such in Ireland probably because the food industry is so dependent on exports.

The Irish food processing industry has seen dramatic changes in its structure in the last generation. While output had grown by a third between 1973 and 1980, and by a further 48% between 1980 and 1989, employment in the food industry had fallen, after an initial rise in the early 1970s. This fall in manufacturing employment was due to increased capitalization and subcontraction of some functions such as milk collection to the service industries.

Value added in the food industry showed healthy growth and the share of value added in gross output increased from 19% in 1977 to 27% in 1987. The structure of value-added has also changed with a decreasing share going to wages and salaries, parallel with the increased capitalization.

The total number of food plants has fallen but this fall was concentrated in the smaller (< 30 employees) and larger (> 200 employees) plant sizes. Plants employing between 30 and 200 people increased in number.

Only 7.8% of food plants were foreign owned in 1988, but these produced 26.7% of the output and accounted for 20.5% of employment. There was a considerable degree of instability in the industry with many new firms starting and many others closing down.

In the Single Market there will be considerable pressure to increase scale of activities to remain competitive. Irish firms are "internationalizing" their activities and have been particularly active in taking over food firms in Britain, though their present scale still renders them too small to be major forces on the European food markets.

The larger Irish companies deliver their own products to the supermarket chains. They also are prominent in the distribution of imported foods, particularly those requiring chilled distribution.

There are also a considerable number of specialist distributors. These firms are faced with a growing market for imported food but also face difficulties in that the large supermarket groups may, in the Single Market, buy directly from manufacturing companies abroad or from a foreign based distributor. Their margins are likely to be squeezed.

Role of Co-Operatives in Agricultural Marketing

Agricultural co-operatives are dominant in Ireland in the following activities (1) sale of cattle and sheep in public auctions (2) purchase and processing of milk and subsequent marketing at home and on export markets of dairy produce (3) processing and marketing of pig meat (4) purchase of grain, manufacture of animal feedstuffs and sale to farmers of grain and animal feed.

Co-operatives are also involved in, but do not dominate, the following activities (1) processing of beef and lamb (2) fresh vegetable growing (3) fish farming (4) pig meat production.

The most powerful co-operatives are those known particularly for their dairy processing, such as Kerry Foods, Avonmore Foods, Dairygold Co-operative, Waterford Foods and Golden Vale. Four of these firms however now have also important meat processing activities. All are involved in addition in supply of grain, animal feed and fertilizer to farmers.

Four of these five co-operatives have recently altered constitutions in such a way that the percentage of the company owned by suppliers may soon fall below 50% and the co-operative character may disappear in time.

Irish co-operatives had considerable difficulty in raising capital from members for expansion and investment. Capital from the Irish banking sector was extremely expensive. Several of the major cooperatives decided about four years ago that the only source of capital available was on the capital markets and converted their companies into public limited companies (plc).

They then raised capital on the stock exchange through the issue of shares, while retaining control of the enterprise in the co-operative. The percentage of the shares held by the co-operative has steadily declined over the period since, as the companies have embarked on a program of takeovers and internal growth. It is possible that in the near future the percentage of shares held by the co-operatives will fall below 50% in some of these companies.

One particular co-operative is worth mention for other reasons. This is An Bord Bainne Co-op, originally a state export board for dairy products. Since 1973 it has operated as a co-operative, with shares held by the dairy processing companies, which are themselves mostly co-operatives. An Bord Bainne does not trade in the home market, except to license its brand "Kerrygold" to some of its individual shareholders, for sale of butter.

Although no longer having any monopoly powers, An Bord Bainne is still the dominant exporter of Irish dairy products, particularly of butter and powders. Its "Kerrygold" brand is one of the few internationally known brands originating in Ireland and it is now the brand market leader for butter on the German market.

An Bord Bainne is active in trying to diversify its product portfolio. It is in close contact with the market and regularly seeks to develop new markets and products, which it then hopes to get its members to supply. The larger manufacturing co-operatives have themselves invested heavily recently in marketing and product development and are active in directly marketing their own new products at home and abroad. An Bord Bainne therefore has to fight to maintain its share of Irish dairy exports, particularly of value added and new products. As part of its development and in order to give it a competitive edge, An Bord Bainne has purchased dairy packaging and distribution companies in several markets, including U.K., U.S.A. and Belgium.

Vertical Marketing Systems

At present vertical marketing systems are not a feature of agricultural marketing in Ireland. However there have been several developments in the recent past, (e.g., the requirement in U.K. law for due diligence on the part of food retailers to ensure the quality of the food they sell), which may lead to much closer involvement between the different links in the marketing chain and to the traceability of the food product to its farm source. CBF, the meat promotional agency has introduced a pilot quality guarantee scheme which will have these characteristics. So has one of the smaller Irish supermarket groups, Superquinn. It is also essential to provide assurances of quality and standards in the growing, but still small, organic foods market.

In general quality guarantee schemes, which involve quality control at all stages from farm to market and traceability back to the farm, are seen as important building blocks in convincing European consumers that the Irish food products they buy are pure, free from antibiotic and other chemical residues and are produced in a "green" environment.

PRODUCT, PRICE, PROMOTION AND DISTRIBUTION

Product

The narrowness of the product range emanating from the Irish agricultural and food industries is seen as a weakness. There is too great an emphasis on production of commodities and of sale of commodities to the Intervention agencies. This criticism applies particularly to the beef and dairy sectors, which together account for almost 80% of agricultural output but a considerably smaller percentage of employment and value added.

The product mix in both of these sectors is heavily influenced by the seasonal pattern of calving and milk production, which is itself related to the predominant farm management system–the feeding of grass. The Irish climate is particularly well suited to grass growth. The production of cereals on the other hand is limited for climatic reasons but also because of soil types. Cereals in Ireland tend to be more expensive than in other parts of the Community while grass is cheaper and more plentiful.

Accordingly the cheapest system of milk and beef production on farms involves maximum use of grass and preserved grass and minimum use of cereals. This leads to a highly seasonal production system with cows calving in early Spring, milk production concentrated in Spring and Summer months and little milk availability between November and February when grass is not generally available. (For the liquid milk market which accounts for only 10% of milk utilization, special incentives are offered to farmers to produce milk to meet these market demands.)

A similar story applies in beef. The bulk of the raw material supply for beef comes from the dairy herds, which provide calves on a highly seasonal basis. These animals are generally reaching maturity at 2 1/2 years, i.e., in the Autumn. They are sold then at the end of the grass growing season, before they need to be fed expensive concentrate feeds.

While these production systems have the merit of using a locally available cheap raw material to the maximum degree, they do cause problems for processors in marketing. The product mix available to the dairy industry is limited substantially to products which are storable and which have a long shelf life. The most modern dairy products and those offering the best margins to processors and the greatest prospects for growth, tend to be those such as yoghurt, soft cheese and fromage frais, which have a limited shelf life. Even with storable products, the technical characteristics of the products change with the seasons making delivery of a consistent product to the market place difficult.

Similar considerations apply in beef. Because of the highly seasonal slaughtering pattern (40% in the period October to December) it has been difficult for meat trading companies to enter long term arrangements for supplying large quantities of beef to supermarkets.

Despite the difficulties in the dominant dairy and beef sectors there has been a healthy flow of new innovative food products from the Irish food processing industry. The poultry industry has been particularly innovative and there has been a steady development in the number and range of further processed poultry products. There has also been a healthy development of production of ready meals from firms in other sectors.

Price

In the commodity markets, Irish producers have historically been price takers, with little influence on the overall markets. This is clearly illustrated by the changes in skim milk powder prices during the period 1988 to 1990. Irish wholesale prices during that period reacted more strongly to the initial increase in world market prices than did those of other Member States of the Community and similarly reacted to the decline in prices when it came (Pitts, 1990).

Some Irish companies have been endeavoring to break free of this dependence on fluctuating world market prices, by catering more for industrial markets for ingredients. These markets require specific functional characteristics, and fluctuate less in price. They also offer an opportunity to create secure long term markets without spending millions on advertising and branding. They are technology rather than marketing dependent.

Kerry Foods is a good example of this development. This company was formerly a substantial exporter of commodity casein to the USA. It has within a few years become one of the major world players in the dairy ingredients business, particularly since its takeover of the ingredients business of the U.S. conglomerate, Beatrice Foods.

In the area of consumer products an alternative, but in some ways similar, policy has been embarked upon. Firms lack the resources in general to compete directly with multi national brands. They have therefore cultivated the supermarket groups on export markets (particularly in the U.K.) and contracted to produce a substantial volume of "own label" products for Sainsbury's, Tesco, Marks and Spencer, etc. The products are particularly in the ready meals area. One of the largest manufacturer of frozen pizza in Europe is a young Irish company, Poldy's.

There are a small number of successful Irish food and drink brands, which have been able to maintain and even increase market share without cutting relative prices. Two examples are Bailey's Irish Cream liqueur worldwide and Kerrygold butter in Germany.

Promotion

There are two State financed promotional bodies, as described earlier, for meat and for fish, which promote the generic consumption of these products at home and support the export efforts of Irish firms in these sectors. There is also a body, the National Dairy Council, which promotes generically the consumption of dairy products. This body is supported by the dairy industry and by the EC Commission. An Bord Glas, the Horticultural Development Board, while primarily active in stimulating production of horticultural crops in Ireland, is also involved to a small degree in promotion of consumption of fruit and vegetables.

Individual companies active in the home market are also active in promoting their own companies and brands abroad. As stated above, the extent of brand activity by Irish companies outside the State is limited by their small scale.

The most important medium for promotion of food products on the Irish market is television, principally RTE, the national station, but there is also substantial promotion on UTV, the Northern Ireland arm of the British independent television network, which is widely viewed also in the Republic of Ireland.

Distribution and Logistics

Most large firms distribute their own products direct to the supermarket chains. Daily or even twice daily deliveries are required for fresh products, while weekly deliveries apply in the case of non-perishable, slow moving, items. Deliveries in rural areas are less frequent.

In the case of smaller firms or foreign firms, whose market volume in Ireland is small, distribution to the supermarket groups and to the wholesale or cash and carry outlets is made by the specialized distributors referred to earlier or in some cases by the larger Irish companies, on contract.

Many of the smaller shops and particularly those in rural areas obtain their supplies of non perishable products from wholesalers or cash and carry firms. Frequently the products are collected by the retailer himself.

Distribution to export markets can be a major problem because of the perishability of many products and the necessity to cross at least one stretch of sea. Most distribution is by road. Efficient distribution therefore depends on a good road network both here and in the U.K., an efficient port operation with frequent sailings to minimize delays, and minimum customs formalities. While there have been considerable improvements in recent years, the logistics of delivering Ireland's very considerable food export volume are a continuing problem, particularly to Continental Europe.

EFFICIENCY OF SYSTEM

Prices Received by Farmer

We now try to assess the performance of the system in delivering high prices to the farmer. Data on this area are deficient. The Irish

Farmers' Association has published data which show the declining share of the pound spent by the consumer which accrues to the farmer. However these data do not take account of the increasing range of other services which the consumer is now getting with her product compared with a generation ago.

One area where some realistic international comparisons are possible is in the dairy area. With a common target price for milk of defined quality and a common intervention price for butter and powder, it is possible to compare accurately the ability of the Irish marketing system with those in other countries to return a high milk price. In this case, we can say, from the analyses carried out, (e.g., Pitts, 1990) that, although the milk price in Ireland is the lowest in the Community in most years, the system works well. High world market prices are reflected in high milk prices and vice versa. The differences between Ireland's milk price and those elsewhere can be satisfactorily explained.

One possible reason why there is effective transfer of prices back to the farmer in the dairy industry is that all stages in the marketing chain are under co-operative control. Elsewhere only in the pig meat industry do similar conditions apply and there only recently.

In other sectors, the farming organizations endeavor when negotiating with processors to obtain some of the benefits of good international prices and to mitigate the effects of bad prices. They are reasonably effective in protecting their members who grow sugar and cereals.

There are some doubts about the effectiveness of the marketing system in the beef and lamb sectors in communicating market requirements back to the farmer and in reflecting accurately the fluctuations in world or European prices.

Competitiveness

The Irish agricultural and food industries have been particularly concerned in recent years with their competitiveness. The emphasis placed on the food industry as an engine of future economic growth is one reason for this concern. Another is the coming of the European Single Market, with the promise of increased competition. Recent experience would seem to indicate that the Irish agri-food industry is indeed competitive. It possesses factor cost advantages in both dairying and beef production, its two most important sectors.

The output of the food industry increased in volume by 31% between 1985 and 1990. Exports also increased substantially. The European market for food is growing by only about 1% per annum. The market share held by the Irish agri-business sector has therefore been increasing in recent years, implying that it is competitive and has little to fear in the Single Market, despite the problems of product mix, dependency on intervention, seasonality of production and relatively small scale of operation.

REFERENCES

Attwood Research of Ireland, 1992. Attwood Consumer Pael Data, Dublin.
European Commission, 1991. The Agricultural Situation in the Community, Brussels, Luxembourg.
Government Publications, 1989. National Development Plan, Dublin.
O'Neill F., 1988. Dietary Concerns of Irish Consumers, Socio-Economic Research Series No 10, Dublin, Teagasc.
Pitts, Eamonn, 1990. European Producer Milk Prices in 1988 and 1989, Information Update Series No. 40, Dublin, Teagasc.

Agricultural Marketing in Belgium and the Netherlands

Matthew Meulenberg
Jacques Viaene

SUMMARY. Agriculture in Belgium and the Netherlands has a strong export tradition and has been market oriented for a long time. In this article agricultural marketing in Belgium and the Netherlands is analyzed on the basis of the concepts structure, conduct and performance. In our review of market structure attention is paid to the structure of agriculture, the food consumer, food retailing, government policies, competition and marketing channels. Afterwards market conduct with respect to product, price, promotion and distribution is discussed. Finally some qualitative observations are made on marketing performance.

It is concluded that agricultural marketing policies in Belgium and the Netherlands are increasingly focusing on value added to the agricultural product. As a result vertical marketing systems/food chain and marketing management become familiar concepts to agricultural marketing.

INTRODUCTION

Dutch and Belgian agriculture by tradition is market oriented. In particular Dutch agriculture is export oriented. The attractive geo-

---

Matthew Meulenberg is affiliated with the Agricultural University Wageningen. Jacques Viaene is affiliated with the State University of Gent.

[Haworth co-indexing entry note]: "Agricultural Marketing in Belgium and the Netherlands." Meulenberg, Matthew and Jacques Viaene. Co-published simultaneously in the *Journal of International Food & Agribusiness Marketing* (The Haworth Press, Inc.) Vol. 5, No. 3/4, 1993, pp. 141-161; and: *Food and Agribusiness Marketing in Europe* (ed: Matthew Meulenberg), The Haworth Press, Inc., 1993, pp. 141-161. Multiple copies of this article/chapter may be purchased from The Haworth Document Delivery Center [1-800-3-HAWORTH; 9:00 a.m. - 5:00 p.m. (EST)].

graphic situation of both countries between urbanized regions of Germany, Great Britain and France, a mercantile tradition, and an open market policy of governments have stimulated market orientation.

In the first half of this century agricultural marketing in both countries primarily was concerned with the effectiveness and efficiency of marketing functions and marketing institutions.

Problems and policies of agricultural marketing have changed during the past thirty years, amongst others because of changing consumers, of changing market structure, and–during the past fifteen years–because of societal concern about the physical environment. As a result marketing management has become relevant in agriculture too.

Not only changes in the socio-economical environment of agriculture and agribusiness, but also endogenous changes in agriculture and agribusiness themselves, such as new production technologies and changing agribusiness structures have increased the relevance of marketing in agriculture.

Agricultural marketing in Belgium and the Netherlands will be discussed in this article, in particular present developments. We will base our analysis on the concepts Structure, Conduct, Performance (Bain, 1956; 1959). The article is organized as follows. Firstly market structure is analyzed by discussing successively consumers, competitors, food retail and government policies. The section on market structure ends up with an analysis of marketing channels and marketing institutions. Subsequently marketing conduct is analyzed with respect to marketing instruments and marketing policies. Finally, the performance of agricultural marketing is evaluated briefly.

MARKET STRUCTURE OF BELGIAN AND DUTCH AGRICULTURE

Some Facts About Belgian and Dutch Agriculture

Agriculture is no longer a dominating sector of the Belgian and Dutch economy: in 1987 agricultural income as a percentage of

National Income amounted in Belgium and the Netherlands to respectively 1.9% and 3.9%. Of total labor force 2.5% and 5.4% were employed in Belgian respectively Dutch agriculture (LEI-DLO, 1989) and an additional 3.2% respectively 3.4% of total labor force were employed in food industry of both countries (LEI-DLO, 1989; NIS, 1991).

Agriculture and agribusiness contribute substantially to national exports: of total Dutch exports in 1990 about 24% was of agricultural/agribusiness origin (LEI-DLO, 1991). Livestock production and horticulture are the most important sectors of Dutch agriculture (Table 1). The livestock sector is transforming feed grains, tapioca and other feedstuffs into protein rich animal food products, in particular milk and pigmeat. Substantial quantities of feedstuffs are imported and livestock products are exported. This marketing policy of Belgian and Dutch agriculture has benefitted from the convenient geographic situation of these countries between population centers of Western Europe: Germany, Great Britain and France.

Horticulture is very important nowadays, realizing in 1990, 23% of Belgian and 30% of Dutch agricultural output (Table 1). Fresh vegetables are mainly being produced in glasshouses. Horticulture has a long tradition in the western part of the countries, near the sea coast. It has expanded substantially since the fifties because of increasing demand for fresh produce.

The relative importance of the three main production sectors (arable land, livestock and horticulture) is quite similar in both countries. Within Belgium, arable production is dominating in the Southern region, while the Northern region is specializing in livestock and horticulture.

The Food Consumer

Food consumption in Belgium and the Netherlands amounted to 17.1% and 18% of total private consumption in 1990 (LEI-DLO, 1992; NIS, 1991). Total food expenditure is increasing modestly and somewhat less than total consumption, the Dutch index of domestic consumption of food and luxuries for instance increased by 13% in volume in the period 1980-1990, while the index of total consumption increased by 16% in volume during the same period.

While total food consumption has increased to a limited extent,

TABLE 1. Value of agricultural production in the Netherlands and Belgium and number of farms in 1990.

	the Netherlands			Belgium		
	billion ECU	%	number farms	billion ECU	%	number farms
Arable production(3)	1.4	8.6	12 590	0.8	13.8	11 425
of which grain	0.2	1.2		0.3	5.1	
sugarbeets	0.4	2.4		0.3	4.6	
potatoes	0.7	4.3		0.2	3.0	
others	0.1	0.7		0.06	1.1	
Livestock	9.7	59.1	59 968	3.6	63.3	41 425
milk	3.8	23.2		0.9	16.0	
meat	4.8	29.3		2.4	41.4	
poultry,eggs	1.1	6.7		0.3	5.9	
Horticulture	4.9	29.9	21 555	1.3	22.9	7 332
vegetables	1.8	11.0		0.7	12.4	
fruit	0.3	1.8		0.2	4.1	
ornamentals	2.8	17.1		0.4	6.4	
Others	0.4	2.4	4 085			
Total	16.4	100.0	98 198(1)	5.8	100.0	60 182(2)

1 ECU = fl. 2.32   1 ECU = 42 BF
(1) exclusive of 26 700 part time farmers   (2) exclusive of 26 998 part time farmers
(3) in the Netherlands, Arable production in 90/91

Source: LEI-DLO, 1991, 1992; NIS, 1991.

the composition of the food basket has changed substantially. In both countries per capita consumption of animal proteins increased during the past ten years, due to rising consumption of poultry meat, pigmeat and cheese. The growth of per capita consumption of fresh vegetables and poultry meat is in conformity with the generally observed increasing preference for food with low fat content. Per capita pigmeat consumption increased because of relatively low prices, while the increase of per capita cheese consumption may be caused by consumers' desire for tasty food and for variety in food. It is interesting to notice that the decrease of per capita food consumption relates to products which had in 1980 a comparatively high per capita consumption in the respective country: bread, potatoes, butter and beef in Belgium and milk, margarine and sugar in the Netherlands.

Various demographic changes, familiar to Western societies, influence the development of food marketing: minor growth, or even stagnation of total population, the ageing of the population and decreasing household size. While Dutch population is expected to increase just from 14.9 million to 15.9 million in the period 1990-2000, Belgian population is forecasted to decrease, from 9.8 million to 9.6 million during the same period. The proportion of people being 60 years and older did increase from 15.6% to 17.5% in the Netherlands during the period 1980-1991, and is forecasted to increase from 20% to 25% in Belgium over the period 1985-2010. It is expected that one person-households will increase from 23% in 1980 to 33% in 2010 in Belgium and from 21.4% in 1983 to 39% in 2000 in the Netherlands (AGB, 1992).

Belgian food consumers are more quality conscious than Dutch consumers (Steenkamp, 1992), for instance, Belgians consumed more high quality food like butter, beef/veal and fresh vegetables than Dutchmen in 1990 (Table 2). Nevertheless there is much similarity between Belgian and Dutch consumers in the product attributes which influence their food choice: attributes determining food choice of Belgian food consumers are freshness, taste and ease to prepare (Huygebaert a.o., 1987); taste was considered the most important of thirteen food attributes by Dutch consumers (Steenkamp a.o., 1986). In other research naturalness, healthiness and absence of noxious additives were reported to be the most important

TABLE 2. Per capita food consumption in the Netherlands and Belgium, 1980 and 1990, kg/capita.

	the Netherlands		Belgium	
	1980	1990	1980	1990
Bread	57.4	59.8	75.4	69.3
Potatoes	83.0	87.0	101.0	91.1
Sugar (basis: white sugar)	41.9	37.4	34.8	42.3
Fruit	67.8	73.0	77.9	58.2
Fresh vegetables	53.2	63.1	59.1	92.3
Milk and milk products	137.1	131.3	93.4	93.6
Butter	3.6	3.5	8.6	7.7
Margarine	12.6	9.8	11.4	12.5
Cheese	13.5	15.1	11.7	15.8
Beef and veal	19.1	18.9	28.0	19.8
Pig meat	39.5	44.0	41.2	45.8
Poultry meat	8.9	17.2	13.0	17.2

Source: LEI-DLO, 1991; NIS.

food attributes for Belgian and Dutch food consumers (Steenkamp, 1992). So it is not surprising that Belgian food consumers in a survey of 1986 expected a consumption increase for low fat milk, low fat cheese, whole wheat bread, fresh vegetables and fresh fruit and a decrease for fat meat, sugar, white bread, butter and cream (Huyghebaert a.o., 1987). Health considerations seem to be the dominating criterion for consumers choosing organic food (Oude Ophuis, 1991).

These developments in consumer behavior demonstrate the need for consumer orientation in agricultural marketing, and have brought product development and promotion in the center of agricultural marketing.

CHANGES IN RETAILING

Like in all Western countries food retailing has shifted from small independents, grocery stores, greengrocers, butchers' shops and bakeries, to large food chains. In Belgium the integrated middle size distribution chains in particular have increased their market share over the period 1980-1990. The distribution or "bolt" law of 1975 regulates the geographical spread of retail outlets. This law aims at optimizing the provision of food products to consumers. As a result the integrated mass distribution could not increase its market share. In the Netherlands no such type of law exists and national food chains increased their share in total food retailing from 21% in 1970 to 47% in 1990 (De Jong, 1989). Dutch integrated distribution realized a market share of 64% in 1990. Integrated mass distribution increased its market share from 37% in 1985 to 41% in 1990 (Table 3).

Developments in food retailing in Belgium and the Netherlands can be characterized by concentration, specialization and internationalization. The market share of the two biggest food retail groups in Belgium has increased from 31% to 41% over the period 1982-1989. Specializing on selling fresh food is becoming more important and food retailing is internationalizing both by penetration of foreign retail companies in the Belgian food market, like the British company Marks and Spencer and the German company Aldi, and by foreign subsidiaries of Belgian retail companies, like

TABLE 3. Shifts in food retailing sector in the Netherlands and Belgium.

Class	Description	the Netherlands [1]		Belgium	
		1985	1990	1980	1990
F1	Integrated mass distribution	37	41	47	47
F2I	Integrated middle distribution	26	23	7	14
F2NI	Not integrated middle distribution* (>400 m²)	24	26	20	25
F3	Other self service stores (<400 m²)* and traditional stores	13	10	26	14
	Total	100	100	100	100

* in the Netherlands, division F2NI / F3 at turn-over of fl. 1.5 million

Note: Classification in the Netherlands and Belgium is not completely identical.

Source: Belgian Committee for distribution; Nielsen (the Netherlands)

[1] Figures refer to general food retailing.

"Food Lion" of the Belgian chain Delhaize in the U.S.A. In the Netherlands the holding company Ahold Ltd, through its food chain "Albert Heijn" and its majority share in a wholesale company which runs a voluntary chain, commands a market share in general food retailing of 37%. German food retail chains, like Aldi and Tengelmann, have penetrated in Dutch food retailing. Small food retailers try to survive by specializing, by joining voluntary chains or by participating in franchise operations. However, their number is decreasing. The following developments have a great impact on market behavior of food chains in Belgium and the Netherlands:

- retail chains develop strong images and focus on well defined target groups, by specific assortment, product quality, services and pricing,
- private labeling will become even more important yet,
- efficiency improvement–in particular logistical efficiency–is important; logistical planning models, scanning operations and Electronic Data Interchange more and more contribute to efficiency improvement,
- cooperation and coordination between retail chains will increase, in particular by franchising and by alliances.

As a result, marketers of food and agricultural products have to deal with food chains which have substantial purchasing power. They have to build up a competitive edge by appropriate marketing policies. Such policies not only require specific product quality and discounts, but also cooperation on advertising and logistical service. They have stimulated marketing management in agricultural marketing and have changed market structure of agriculture and food industry.

GOVERNMENT POLICIES

Government is playing an important role in agricultural markets. Classical reasons for government market interference are food security, farmers' welfare, economic importance of agriculture, respectively health and environmental considerations. Except during world war periods, food security has not been an important motive in Belgium and the Netherlands. Export orientation and limited importance of arable farming resulted in a non-protectionist food policy in these countries.

Belgian and Dutch governmental agricultural policy has to operate within the rules of the CAP. In addition governments have subsidized some types of investments in order to improve farming efficiency. Government support to farmers has been focused in particular on the improvement of farming quality by research, extension and education.

A number of factors, like dissatisfaction with CAP, environmental problems, the smaller economic importance of agriculture, and budget problems diminish willingness of Belgian and Dutch politicians to support agriculture. So, there is a tendency nowadays to privatize a number of public services to farmers, e.g., agricultural extension services.

Like in other countries, government is involved in agriculture by protecting consumers and environment: food and drugs laws, antitrust legislation, legislation on the use of insecticides, pesticides and drugs and, increasingly, environmental legislation which put side conditions on agricultural production and marketing. In the future agriculture in Belgium and the Netherlands will have to rely more on marketing potential and less on government support.

Changes in Competition

Competition in agricultural and food markets has increased because of the EC. In evaluating the competitive strength of Belgian and Dutch agriculture, Porter's criteria "factor conditions, demand conditions, related and supporting industries, respectively firm strategy, structure and rivalry" (Porter, 1990) seem relevant. With respect to factor conditions, it can be argued that the created factors, i.e., skilled farmers and good allocation of farm land, are more important in giving agriculture a competitive edge than natural advantages such as to climate, soil and water. As far as demand conditions are concerned, it looks as if Belgian consumers are more quality conscious, and Dutch consumers more price conscious. In this respect consumers' attitude, at least in the Netherlands, does not seem instrumental in strengthening the competitive position of agriculture. Belgian and Dutch agriculture seem to have a competitive advantage on the criterion 'Related and supporting industries.' Agriculture, in particular horticulture, in Belgium and the Netherlands is concentrated in specific areas, where there is a good infrastructure of supporting industries and trading companies, which deliver specialized products and services to farmers. Various producer networks exchange experiences and opinions on a regular basis. Firm strategy, structure and rivalry in domestic market, the fourth criterion of Porter, are in particular relevant to agribusiness and food companies, because of the open EC market. Because of strong export orientation, Belgian and Dutch agribusiness are more used to operating in competitive markets than some of their colleagues in the EC. Porter mentions two additional factors being of importance for competitiveness, namely government and chance. Governments have paid ample attention to the improvement of farmers' skills and of infrastructure, which has strengthened competitiveness. Competitive position of Belgian food industry could be strengthened by improving technology and workers' skill, respectively by increasing company size. In the Netherlands also a large number of companies are too small to compete effectively in international markets (Table 4).

Belgian and Dutch agriculture and food industry have built up a strong export position. Belgium has become a net-exporter of food products since 1983, while the Netherlands has been a net exporter of food products for a long time. Belgian and Dutch food and agricultural products are mainly exported to EC countries, in particular Germany, France and the United Kingdom. While France is the most important export market for Belgian food products, Germany is the dominating export market for Dutch agricultural and food products. Dutch agriculture/horticulture and food industry also export substantial quantities of products, like flower bulbs, to third countries out of the EC. Product quality and services have become very important as a competitive weapon.

A competitive disadvantage of Dutch agriculture, and to some extent of Belgian agriculture also, are the costs of environmental problems, caused by the manure from intensive methods of animal husbandry and by the intensive use of fertilizers and pesticides.

TABLE 4. Turn-over and employment in food industry in the Netherlands (1989) and Belgium (1990).

NACE	Subsector	the Netherlands				Belgium			
		Turn-over b ECU	%	Employment x1000	%	Turn-over b ECU	%	Employment x1000	%
411	Vegetable and animal oils and fats	1.9	6		a	1.1	6	2	2
412	Slaughterhouses, preparing and processing of meat	6.8	22	22	15	2.0	11	12	14
413	Dairy products	5.4	18	19	13	2.9	16	8	8
414	Fruit and vegetable processing and preserving	1.1	4	10	7 b	1.0	5	4	5
415	Preparing and preserving of fish	0.4	1		b	0.2	1	1	1
416	Flour	0.6	2	14	9 c	0.7	4	2	2
417,418	Starch and pasta	0.7	2		a	0.4	2	1	1
419	Bread and biscuits	2.0	7	44	29	1.7	9	26	29
420	Sugar	0.7	2	3	2	0.7	5	2	3
421	Cocoa, chocolate and sugar confectionery	1.4	5	7	5	1.0	4	9	10
422	Feed	4.4	15		c	2.2	12	4	4
423	Other food	2.5	8	21	14 a	1.5	8	6	7
424-426	Alcohol, spirits and wine	0.4	1	11	7 d	0.3	2	1	1
427	Beer and malt	1.4	5		d	1.8	10	9	10
428	Water and soft drinks	0.6	2		d	0.7	4	3	3
	Total	30.3	100	151	100	18.2	100	91	100

a NACE 411, 417 en 418 included in NACE 423
b NACE 415 included in NACE 414
c NACE 422 included in NACE 416
d NACE 427 and 428 included in NACE 424-426

Source: C.R.B. - Bijzondere raadgevende commissie voor de voeding; C.B.S.

MARKETING CHANNELS
AND MARKETING INSTITUTIONS

The structure of agricultural marketing channels is changing. The functions to be performed in the channel, the institutions of the marketing channel and the relationships between these institutions are changing.

Farms, being mainly family farms, have specialized, for instance into fruit growers, growers of fresh vegetables, poultry farmers, pig farmers and dairy farmers. Typically, agricultural products are marketed by wholesale companies and/or processing industries, both private and co-operative. Direct selling from farm to consumers is of minor importance.

Agribusiness and food business increasingly contribute to the marketing of agricultural products by processing, packing and offering services to the final customer. Food industry realized 3.2% of the gross national product in the Netherlands and 3.2% in Belgium in 1990.

There are specific differences in the importance of various sectors of the food industry between both countries: in Belgium, breweries are more significant since by tradition beer consumption is of more importance; in the Netherlands processing of cocoa-beans is more important because of the historical relationships with tropical production areas. In both countries there are a great many small food companies as yet. In Belgium for instance 82% of the food companies, being responsible for 19% of the employment of the sector, do have less than 10 employees. Concentration and internationalization are central tendencies in the food industry. In Belgian food industry a few large national companies are market leaders, like Vandermoortele for oils, Van den Broecke for processed potatoes, Leonidas for pralines, Spadel for mineral water and Interbrew for beer. In other fields foreign companies are market leaders. In the Netherlands, multinational food companies are leading in various food markets, like Unilever in margarine, D.E./Sarah Lee in the coffee market, while three national co-operatives dominate the dairy market. Increasing competition of multinationals in the large European food market, and concentration in food retailing are main challenges to the food industry. The aim of creating added value by processing agricultural products stimulates attention for product quality, amongst others by integrated quality control and branding.

Logistics is becoming more important in order to increase customer service and to improve efficiency.

Belgian and Dutch wholesalers of agricultural and food products have lost market share, like for instance in marketing pigs, cheese and groceries, to forward and backward integrating industries and retail chains. They have maintained a strong position in marketing of perishables like flowers, fresh vegetables and potatoes.

Co-operatives are very important in Dutch agriculture and are of substantial importance in Belgian agriculture (Table 5). They are in particular important for perishable products which cannot be stored at the farm, like milk and fresh horticultural products, and in agricultural sectors having an important export share.

Many co-operatives started out as local or regional businesses at the end of the nineteenth century. They became more market oriented in order to suit the needs of the market. Market orientation stimulated a development towards large co-operatives; in the Netherlands three co-operative companies account for more than 80% of Dutch milk supply, two co-operative auctions dominate the Dutch flower market and one co-operative the potato processing for industrial purposes. By specific organizational structures, like a limited company, whose shares are owned by the co-operative union, co-operatives try to improve effectiveness of decision making and of marketing policies.

Contract farming is important for broilers and for vegetables for canning purposes. In Belgium contract farming has a market share of 95% for veal, 90% for broilers, 70% for eggs and 60% for pig meat–feed companies and slaughterhouses being the main contractors–flax is almost exclusively produced under contract and potatoes for about 50%.

Technical markets have become less important in agricultural marketing. They are still important for livestock, and in particular for fresh horticultural products. In the Netherlands there are two futures markets, respectively for potatoes and live pigs.

Being introduced in 1887, auctions have gradually become the dominating marketing institution in Dutch horticulture (Meulenberg, 1989). They are very important in Belgian horticulture too. Dutch auctions have expanded activities from price discovery and product assembly into other marketing activities, such as minimum

TABLE 5. Processing and selling co-operatives in the Netherlands in 1985 and 1990: number, turnover (in billion guilders), and market share (in percentage).

Year	Number '85	'90	Turnover '85	'90	Market Share '85	'90
Dairy	22	18	11.2	10.5	86	84
Livestock	2	2	3.1	3.2	26	24
Meat					16	16
Poultry	2	2	0.3	0.3	27	21
Sugar	1	1	0.9	1.0	63	63
Potato						
processing[1])	1	1	0.8	1.0	100	100
processing[2])	2	3	0.3	0.5	28	31
Fruit					75	77
and	41	28	3.1	3.9		
Vegetables					80	70
Flowers					92	95
and	12	8	3.3	4.7		
Plants					67	72
Eggs	3	2	0.3	0.2	18	16
Ware potato					31	45
and	9	9	0.5	0.8		
seed potato					47	61
Flower bulbs	1	1	0.3	0.4	47	50

1) For industrial application;
2) For human consumption, like chips, crisps.

Source : Coöperatie (1991)

pricing schemes, promotion, logistics, and product policy. Dutch auctions cooperate in master organizations which coordinate activities, like national minimum price schemes and promotional activities. Some retail chains are critical of the auction system, since daily purchasing through auctions does not fit to retail sales planning. Co-operative auctions have developed additional selling operations, such as brokerage operations in pot plants and auctioneering for delivery at a future time period.

In Dutch agriculture commodity boards exist in many sectors, like milk and dairy products, meat and meat products, and ornamentals. They do not engage in buying and selling but are engaged in market research, promotion and technical research for the generic product. The total promotional budget of Dutch agricultural commodity boards amounted to 144.2 million guilders in 1989. Commodity boards were set up in the fifties, when agricultural markets typically existed of a great many small farms and processing companies. At present various agribusiness firms, including co-operatives, have become national or even international companies which run individual marketing programs. As a result agribusiness firms are less in a need for generic marketing programs like those set up by commodity boards.

MARKETING CONDUCT OF BELGIAN AND DUTCH AGRICULTURE

Marketing conduct of Belgian and Dutch agriculture and agribusiness is evolving from performing marketing functions (exchange-, physical- and facilitating functions) to marketing management (organizing the marketing mix in view of customers wants and needs). Clearly there is no strict borderline between these two approaches. In fact, marketing management is integrating the planning of marketing functions in a broader framework. Marketing conduct of Belgian and Dutch agriculture will now be discussed within the framework of the marketing instruments.

Product. Product policy has become a core element of agricultural marketing programs. Product quality and product assortment are very important marketing topics: Dutch pig sector is aiming at leaner meat, the percentage of lean meat increased from 53.2% in

1990 to 53.8% in 1991; dairy industries processed more milk into more profitable products like desserts and cheese (45.4% of Dutch milk supply was processed into cheese in 1989 and 48.6% in 1991); during the period 1989-1991. Dutch production of red peppers, mushrooms and cucumbers has increased more than that of other vegetables (LEI-DLO, 1991, 1992).

Research and development, product innovation and branding have become extremely important in agricultural marketing. Integrated quality control, certification of products and production processes are introduced, amongst others for fresh meat, in order to guarantee product quality. Environmentally friendly food production is expanding, like for instance organic food production and animal friendly production methods. The market for organic food in the strict sense, like bio-dynamic and ecofood is limited yet.

Price. Traditionally, agricultural prices were determined by the interaction of supply and demand at technical markets. In some instances, like fresh horticultural products, this still happens to be the case. In other markets price formation by contract is important, such as in the case of fruit and vegetables for canning purposes and markets for broilers.

In Belgium and the Netherlands marketing margins between retailer and producer prices are increasing because more value is added to the agricultural product. In Belgium producer share in consumer prices decreased from 70% to 58% for beef and from 54% to 42% for pigmeat during the period 1981-1990 (Huyghe, 1991). In the Netherlands the share of agriculture, in total expenditure for food and luxuries, decreased from 41% in 1961 to 28% in 1988, the share of distribution increased from 35% to 45% and the share of processing from 24% to 27% (LEI-DLO, 1992).

Promotion. In agricultural and food marketing promotion is an extremely important instrument for creating consumer awareness and product image. It is increasing in conjunction with product differentiation and competition. Co-operative promotion by food producers and food retail chains is important in marketing agricultural and food products too.

In spite of the growing importance of company brand promotion, collective promotion for agricultural and food products is of importance yet. In Belgium three institutions are engaged in collective promotion for agricultural and food products. The first is ONDAH, established under the direction of the Ministry of Agriculture, which aims at controlling product quality and at the promotion of agricultural products. Since 1984 the private sector is more involved in ONDAH by fund raising and decision making. Since 1990 two marketing offices, in Paris and Bonn, support Belgian exporters of food and agricultural products. The second institution is BDBH (Belgian Office for External Trade), a public institution established in 1948, which is offering trade services for semi-processed and processed export products. The third is Vitabel, the export department of the Federation of the Agri-Food Business, LVN, which supports producers and wholesalers of food products, in particular with respect to foreign trade fairs. The tendency towards regionalization of Belgian collective promotion leads to scattered promotions and promotional impact.

In the Netherlands commodity boards and co-operative master organizations, for instance in fruit and vegetables, are engaged in collective promotion. Collective promotion financed by commodity boards amounted to 144 million guilders in 1989. Determining the appropriate balance between brand promotion by individual companies and generic promotion is a hot issue in Dutch agricultural marketing.

Distribution. Food producers compete for shelf space of retail companies. Purchasing loyalty of retail companies has become very important for Belgian and Dutch agricultural and food producers. Building a direct relationship with food retail chains is instrumental in this respect. As a result distribution policies in agricultural marketing are increasingly concerned with strategic decision making about marketing channels. In various sectors of Belgian and Dutch agriculture vertical marketing systems have emerged: some dairy co-operatives have integrated cheese wholesale companies; some compound feed companies have integrated meat packing companies.

Logistical operations throughout the marketing channel of fresh products such as the marketing channels for fresh meat or pot plants are planned more thoroughly. New information technology, like E.D.I., is helpful in this context.

Transport is particularly important for voluminous perishable products like fresh horticultural products, fresh meat and dairy products. Food producers and middlemen often purchase transport services from specialized transport companies. In the Netherlands 49% of wholesalers in agricultural products and 39% of wholesalers in food products contract out transport in domestic markets; these figures are respectively 77% and 63% for international transport (Nederlands Verbond van de Groothandel, 1991).

PERFORMANCE OF AGRICULTURAL MARKETING IN THE NETHERLANDS

We do not have data at our disposal to assess marketing performance at the sector level, or of some specific marketing companies by methods developed for that purpose (Bonoma and Clark, 1988). So our discussion of marketing performance must remain tentative and of a qualitative type.

Some indication of marketing performance/market orientation might be provided by farmers' income, which, however, not only depends on marketing operations but also on production efficiency of farmers and on farm support by CAP. It appears that family income per family worker in agriculture is higher in the Netherlands than in any other country of the EC: 25,200 ECU in 1988/89 and 31,700 ECU in 1989/90 as compared to respectively 8,200 ECU and 9,400 ECU in the EC on average and respectively 22,300 ECU and 27,100 ECU in Belgium, being second highest (LEI-DLO, 1991, 1992). These figures suggest a fair marketing performance of Belgian and Dutch agriculture, at least relatively speaking.

High prices might suggest monopolies or inefficiencies in marketing operations. However, market prices can be high because of more built in services or higher product quality, or because of CAP measures. Farmers' prices in Belgium and Netherlands in general are low as compared to prices in other EC countries (Eurostat). Prices are in particular low for pigs, poultry and eggs.

Another indicator of marketing performance might be the dynamics of market supply, both in terms of quantity and quality. In fact Belgian and Dutch agricultural markets have been very dynamic during the past ten years. Sales and exports have changed substantially, in particular in horticulture, poultry business and dairy industry (LEI-DLO, 1991).

Quality control has received much attention. In horticulture product innovation has been substantial. In fruit growing new apple-varieties have been introduced. The assortment of vegetables in glasshouses has been extended substantially. The number of new flower varieties has been overwhelming. In arable farming, being a troubled sector today, product innovation to a large extent has been experimental without great success up till now. Production and marketing of organic food, or of products produced in an animal friendly way, is increasing.

CONCLUSION

Marketing agricultural products in Belgium and the Netherlands is increasingly an operation in an environment characterized by a saturated demand, by a changing CAP, an environmentally conscious society and a more critical and better informed consumer. These challenges require marketing policies aiming at more added value and environmentally friendly production methods.

As a result marketing management has become more familiar in agricultural marketing and has stimulated larger firm size in agribusiness. As a result mergers are under way leading to internationally oriented agribusiness companies which have entrenched themselves firmly in the food chain.

LITERATURE

AGB, (1992). Jaargids, Kerncijfers voor markt- en beleidsonderzoek, Dongen.

Bain, J. S. (1959). Industrial Organization, New York, John Wiley and Sons.

Baln, J. S. (1956). Barriers to New Competition, Cambridge, Harvard University Press.

Bertrand, J. M. (1989). Les filières agro-alimentaires. Etude bibliograpbique. Documents de l'I.E.A., n° 25, février 1989, 48 p.

Bertrand, J. M. (1989). Les filières agro-alimentalres. Etude bibliograpbique. Documents de l'I.E.A., n° 25, février 1989, 48 p.

Bonoma, T. V. and B. H. Clark (1988). Marketing Performance Assessment, Harvard Business School Press, Boston, Massachusetts.

Cassady, R., 1967. Auctions and Auctioneering, University of California Press, Berkeley.

Centrale Raad voor het Bedrijfsleven: Bijzondere Raadgevende Commissie voor de Voeding (1991). Verslag over de economische ontwikkeling in de voedingsmiddelensector in 1990 en tijdens het eerste halfjaar van 1991. Brussel, 5 juli 1991, 48 p.

Centrale Raad voor het Bedrijfsleven: Bijzondere Raadgevende Commissie voor de Voeding (1991). De voeding en de dranken in het particulier verbruik. Brussel, 4 November 1991, 23 p.

Commission of the European Communities: Directorate-general for Agriculture (1990). The agri-food business in the Community. CAP working notes, 1990.

De Craene, A. & Viaene, J. (1990). Promotie van de export van voedingsmiddelen in het perspektief van 1992. Seminarie voor AgroMarketing, RUG, in opdracht van het Staatssecretariaat voor Europa 1992, oktober 1990, 98 p.

Gallet, G. & Viaene, J. (1989). De impact van 1992 op de voedingsindustrie. Seminarie voor Agro-Marketing, RUG, in opdracht van het Staatssecretariaat voor Europa 1992, mei 1989, 172 p.

Huyghe, F. (1991). De groot- en detailhandelsmarges voor rund-en varkensvlees. CLEO-schriften, nr. 59, juni 1991, 57 p.

Huyghebaert, A.; Viaene, J. & De Vrieze, M. (1987). Perspectieven van de voedingsconsumptie in Belgie. Diensten voor Programmatie van het Wetenschapsbeleid, Brussel, 1987, 93 p.

Jong, J. G. A. M. de (1989). De functionele plaats van de distributie in beweging: 1970-1990, in: Bunt, J., Dreesmann, A. C. R. and C. Goud, 1989, Dynamiek in de Distributie, Kluwer, Deventer, pp. 3-19.

LEI-DLO (1991). Landbouweconomisch Bericht 1991, Den Haag.

LEI-DLO (1992). Landbouweconomisch Bericht 1992, Den Haag.

Ministerie van Landbouw (1991). Evolutie van de land- en tuinbouweconomie (1990-1991), 29ste verslag voorgelegd door de regering. L.E.I., Brussel, November 1991, 185 p.

Oude Ophuis (1992). De Europese markt voor alternatief geproduceerde voedingsmiddelen. In: Steenkamp, J.E.B.M. (ed.). De Europese consument van voedingsmiddelen in de jaren negentig, Van Gorcum, Assen/Maastricht, pp. 31-47.

Porter, M. (1990). The Competitive Advantage of Nations, MacMillan, 1990.

Rabobank (1991). Cijfers en Trends, Rabobank, Nederland.

Smith, J. (1990). The Community agri-food industry in the single market. Club de Bruxelles, 1990.

Steenkamp, J.E.B.M. (1992). Cross-nationale analyse van het kwaliteitsbewustzijn van consumenten met betrekking tot voedingsmiddelen. In: Steenkamp, J.E.B.M. (ed.). De Europese consument van voedingsmiddelen in de jaren negentig, Van Gorcum, Assen/Maastricht, pp. 127-136.

Steenkamp, J.E.B.M.; Wierenga, B. & Meulenberg, M.T.G. (1986). Kwaliteitsperceptie van voedingsmiddelen. Swoka, Verkennende studie, 40-1, 402-2, Den Haag.

Van Den Bulcke, D. et al. (1988). De Belgische voedingsnijverheid: economische tendensen en technologische perspectieven. Diensten voor Programmatie van het Wetenschapsbeleid, Brussel, 1988, 72 p.

Viaene, J. (1991). Aspects socio-économiques des mutations du complexe agroindustriel. Annales de Gembloux: - 1997. 97: p. 95-113.

Agricultural Marketing in Spain

J. Briz

SUMMARY. The paper deals with the analysis of marketing in the Spanish agrofood sector. The methodology follows the industrial organization paradigm: structure, conduct and performance. It gives a first analysis of market environment, monetary fiscal policies, and consumer's demand.

Market structure shows a dominance of minifundia at farmer and retail level. A concentration process is going on at wholesale and industrial level.

Conduct and performance of agrofood marketing are being studied, with special attention to competitiveness and price evolution.

Finally some observations are made on marketing characteristics in some of the traditional food sectors.

INTRODUCTION

Agricultural marketing in Spain will be analyzed according to the methodology of the industrial organization paradigm (Bain, 1968). The main concept, structure, conduct and performance, is pursued along different dimensions, some of which were discussed in Spain in the early seventies (Briz, 1972).

Many studies deal with problems at farm level in Spain, but only very few pay attention to agricultural marketing. One of the reasons

---

J. Briz is Professor of the Departamento Economia y Ciencias Sociales Agrarias at the Universidad Politecnica de Madrid.

[Haworth co-indexing entry note]: "Agricultural Marketing in Spain." Briz, J. Co-published simultaneously in the *Journal of International Food & Agribusiness Marketing* (The Haworth Press, Inc.) Vol. 5, No. 3/4, 1993, pp. 163-177; and: *Food and Agribusiness Marketing in Europe* (ed: Matthew Meulenberg), The Haworth Press, Inc., 1993, pp. 163-177. Multiple copies of this article/chapter may be purchased from The Haworth Document Delivery Center [1-800-3-HAWORTH; 9:00 a.m. - 5:00 p.m. (EST)].

is the lack of information, the minor interest in this field is another. However, this lack of interest should vanish in the near future as a result of the increasing importance of food industry and distribution sectors. In this paper we want to discuss the whole marketing process from farmer to consumer. However, the difficulty in obtaining information in some areas (behavior or profitability of enterprises) and the lack of space, put restrictions on our exposition.

MARKETING ENVIRONMENT

Over the last decades, a large number of sociopolitical events have conditioned the economic and marketing evolution in the Spanish agricultural sector. The new democratic regime (since 1976) and the integration in the EC (1986) have provoked liberalization of foreign trade and strengthened the incentives for innovations.

Inflation (around 6%) and unemployment (16% in 1990) are the most significant problems which the Spanish economy is facing at the moment. Therefore, public policy is oriented towards monetary and fiscal policies. High rates of interest increase investment costs in the food sector. The high exchange rate of the peseta (Spanish currency) in relation to that of some competitors, diminishes Spain's competitiveness in international markets.

Similar to other countries, agricultural policy is very often in conflict with food policy. Prices of agricultural products at farm level, rising due to market regulations, lead towards growing costs of raw material in food industry. On the other hand, growing competition in markets is lowering prices of final goods, and food industry has been caught between both converging trends.

Food industry in Spain accounted for 17% of industrial turnover, employed more than 400,000 people and contributed 4% of GNP in 1990. During the last decade the rate of growth has been 2.5%, for the last 3 years it has been 3.5%. In comparison, the contribution of agriculture to GNP in 1990 was 4.5%.

Food demand in Spain has a low rate of growth (0.5% per year). Food expenditure in Spain is the most representative index to evaluate it, and this has been very even during recent years (Anuario Distribucion Española, 1990, p. 42).

In contradistinction to the agricultural sector, government policy has been very liberal in regulating other fields: manufacturing industries, distribution and investment activities. Nowadays Spain has one of the most liberal systems of distribution organization, which is one of the reasons for the rapid changes in food retailing in recent years.

Next to that, the agricultural sector follows the evolution of the Common Agricultural Policy (CAP), which is also in a process of profound transformation.

THE MARKET STRUCTURE

During the last decades, the Spanish agrofood sector has experienced profound changes, from a more subsistence oriented agriculture to a modern, commercial one. Although the main marketing functions remain to be performed, substantial modifications have taken place in enterprises, production systems and consumption habits. The structural changes have been stimulated by radical political, economic and social transformations in our society, since the fifties and especially in the eighties. The historical transformation of our markets may be divided into two periods (Briz, 1985). The first one up to 1959, the year of the 'stabilizing plan.' In this period there was a predominance of traditional economy and techniques and of small commercial enterprises in the sector. The second period is from 1959 up to the present. Its main characteristics are the spreading of new activities in organizations, management and technology.

The first period was that of traditional commerce. The pattern is similar to what happened in other countries. It was an incubation period, with great migration movements from rural to urban areas, from agriculture to industry and services. Rapid increase in per capita income, migration to other European countries and massive tourism were all important developments during the early fifties, creating a demand for improved marketing services.

The second period may be indicated as 'mass distribution.' According to Padberg and Thorpe (1974) this stage involves changes in the disposition of the shop itself, and intensification of integration and self-service. The most sophisticated marketing systems are located in the Spanish areas with highest population densities and highest per capita income. The hypermarkets, considered by some people the paradigm of mass distribution, started out in 1973, and in 1990 their number had gone up to 135 (see Table 1). New technology is introduced into marketing companies, which is facilitated by foreign investment. The use of computer programs and even scanners (the 'second commercial revolution after the self-service') has become more common.

Historically, the traditional marketing channel has been farmer-wholesaler-retailer. Government policies in this field were trying to stimulate wholesale markets in producing areas by establishing a national association called MERCORSA and founded in 1970, with capital from farmer cooperatives and government. Today there is a national network of these markets.

The large-scale migration in Spain, from rural to urban areas, has changed the marketing structure. The growing demand for food in the new urban areas required more services, such as transport, storage and financing activities. Under these conditions Central Markets have been the cornerstone in the marketing change. Traditional urban markets are transformed into modern ones, such as hypermarkets. Table 2 shows the relative importance of different sectors at wholesale level.

In big cities the aim was to promote Central markets within a national association named MERCASA (founded in 1966). Control of this association is being exercised by public institutions (central government, local authority) and private sectors. Notwithstanding the managerial and financial problems of the public institutions,

TABLE 1. Evolution of hypermarkets in Spain

Year	1973	1975	1980	1985	1990
Number of places	1	6	31	65	135

TABLE 2. Wholesale trade in Spain (1985)

	Number of enterprises	Number of selling places	%
1 Food and drinks	47.662	18.075	38,7
2 Textiles	10.203	2.516	5,4
3 Pharmacy	10.849	2.303	4,9
4 Durable products	31.240	3.685	7,9
5 Others	30.738	20.086	43,1
TOTAL	130.692	46.665	100

Source: Mercasa and Iresco

their impact has been significant. There is a great diversity of other market organizations, as a result of sector and region specific characteristics.

Auction markets in fruit, vegetables and fish, are traditional, and have been so for centuries. They are located in the North and South of the peninsula.

Forward markets have been functioning in an informal way since the Middle Ages, and nowadays it is common practice to operate in future markets. Especially imported commodities (corn, soybeans) are being traded, through brokers from Chicago, New York and other international boards of trade.

Marketing boards have been efficient only for traditional export products (e.g., citrus board of trade). Their goals were to maintain quality level, marketing coordination and some price policy agreements. With Spain's integration in the EC, their importance is diminishing somewhat because a number of their functions conflict with new EC regulations.

Vertical integration is very low for most Spanish agricultural products, for historical reasons. The highest level of integration exists in the compound-feed, poultry and hog industries. The most important examples of integration are of the horizontal type. Co-operatives are important at farmer level in some specific sectors where economic problems force them to associate in order to survive; wine, olives, fruit and horticulture are some of them (Herrero 1976).

Food industry is in a dynamic structural change. The numbers of enterprises and laborers are decreasing. Value added per enterprise is increasing in some sectors (beer, sugar, canned fish), others (bread, cider, slaughter houses) maintain their levels. There is a rising cost of wages and salaries per person (in particular in the production of beer and liquors). Simultaneously there is a trend of increasing value added per employee, in particular in the production of liquors, beer and sugar.

Although the occurrence of food industries is somewhat higher at the North and East coast of the peninsula, in general they are spread out in all areas. In addition to comparative advantages in the production of raw materials, protectionism made it possible to maintain marginal businesses.

The Spanish retail sector is atomized, and frequently characterized by a destructive competition. Traditional retailers are the most
significant businesses, but modern commercial systems (great mag-
azines, self services, hypermarkets) have increased rapidly in number during the last decades.

Food stores and stores for household products are the most numerous in retail trade (Table 3).

Small enterprises are found most frequently. However, a concentration process is under way. In 1989, 38% of total sales were
controlled by the largest 50 enterprises (Mortes 1990).

The structure of food retailing shows a predominance of traditional, small family businesses, dealing with fresh, perishable products. They represent more than 80% of retail business for bread and
fish, 70% for meat, 60% for fruit and vegetables, and 50% for eggs
and pasteurized milk.

Hypermarkets are taking over market share from traditional
stores, especially in fruit juices, oil, dairy products and dry fruits. It
is one of the most dynamic distribution channels. As we mentioned,
in 1990 there were 135 distribution points, with 1840,000 m^2 of
selling area, and 23,300 employees. There is a significant concentration ratio. Three French enterprises (Pryca, Continente and
Alcampo) control 58% of the selling area.

Wholesaler sponsored voluntary chains are also significant in
food retailing. In 1990 there were 225 enterprises with 10,000 selling points, most of them in the food business. The most important
ones are SPAR, SOGECO and IFA. Problems they have to deal with
are for instance 'buying misbehavior' of retailers and abuses of
wholesalers.

Consumer cooperatives are decreasing in importance, as a consequence of low adaptation to market conditions, old fashioned outlets and low innovation activities. There are about 200 cooperatives,
with 5% of total sales in the sector. The biggest is U.D.A. with 44
members in 1990, and 60% of co-operative sales. Cash and carry
wholesale companies are increasing in number as a result of traditional wholesalers' interest in this type of business, and the restructuring by independent entrepreneurs of some of their warehouses.
In 1990 451 firms had 841 selling points in 'cash and carry' whole-
saling.

TABLE 3. Retail trade in Spain (1985)

	Number of enterprises	Number of selling places	%
1 Food	284.206	256.137	40,4
2 Drugstores	56.882	42.720	6,8
3 Textiles	113.163	129.989	20,6
4 Menaga	80.583	80.042	12,7
5 Others	180.917	123.450	19,5
TOTAL	715.751	632.338	100

Source: Mercasa and Iresco

CONDUCT OF MARKETERS
OF AGROFOOD PRODUCTS

Getting information on market conduct is difficult, especially where the behavior of private entrepreneurs is concerned. However, some authors (Scherer 1970) consider it to be one of the more important characteristics of industrial organization.

The future of the Spanish agrofood sector to a large extent depends on introducing new technologies, but R and D investments are very low. During the period 1980-81 the ratio of expenditures on R and D over Gross Value Added was 0.5 in Food Industry and 2.0 in the total Industrial Sector.

Changes in food demand and growing competition should stimulate marketing and production innovation in public and private sectors. At farm level, research and innovation activities are carried out by official institutions such as the Instituto Nacional de Investigaciones Agrarias (INA).

New technologies in the food industry mainly stem from Switzerland and USA. The multinational Nestlé (from Switzerland) takes care of more than 85% of expenditures in R and D. In the distribution sectors the innovations originate from French multinationals mainly. Studies show that 66% of the food industrial enterprises do not have any activity in R and D. This proportion is lower in agriculture and distribution.

Grading and quality control have been improved during the last decades. Regulations have been provided following international rules (Codex Alimentarius Mundi, FAO and OCDE indications, etc.), and especially the EC practice. Today, there are quality and grading norms for most food products, issued by national government. Adequate controls are provided by public institutions, for foreign as well as domestic markets. Both the Ministry of Agriculture, Food and Fisheries, and the Ministry of Health have responsibilities in this field, and this situation sometimes creates problems of coordination and efficiency. In any case the standards to be introduced are discussed in advance with the entrepreneurs and experts.

Price formation has been following market rules, with government intervention on 'strategic products' in the 'food basket.' However, the EC membership diminished the possibility of pure Government interventions. In this respect, one should remember that the EC has a CAP (Common Agricultural Policy) but not a Common Food Policy. Therefore CAP price policies have been mainly oriented on stimulating or protecting producers, and not on protecting consumers.

Marketing margins in general are larger than in other European countries, at least in relation to quality and services provided. Transport costs are higher also, because of unfavorable geographic conditions. According to IRESCO (1976, p. 53), transport costs in relation to total output were 4% in Spain, 3.1% in France, 3.3% in Belgium and 3.7% in Italy in 1970.

Promotion has traditionally been done by export sectors (citrus, olive oil, wine), and is mainly oriented towards foreign countries. They received financial and professional support from official institutions (Instituto Nacional Fomento Exportaciones, transformed later into Instituto Español de Comercio Exterior). Today the EC regulations are followed in this field.

The domestic market was considered an isolated, captive market for the national producer sector. However, with more competition, as a result of the 'liberalizacion,' new promotional strategies have been put into practice.

Publicity in international and domestic markets for generic products is often focused on characteristics related to geographic origin (Denominacion de Origen). In this manner a national institution (INDO) is sponsoring promotion in generic products (wine, olive oil), besides quality control activities. Professional organizations are involved in these activities too (Caldentey 1987). Certain individual big companies are also involved in promotion, aiming at establishing strong private brands.

A matter of concern of many agricultural and industrial firms is the growing number of private labels. The resulting bargaining power of big food chains has created problems to food industry in some sectors (milk, juices). Multinational companies have been making great efforts in food promotion in recent years. One time they use international programs (usually in American soft drinks, Italian 'pastes'). Another they just mention some relevant activities in favor of environment protection in certain regions, related with the production of some agricultural products.

In 1987 total expenditure on publicity in Spain amounted to 540,000 millions of pesetas, with 28552 trade brands (Alvaro Taboada in Briz, 1990). Newspapers (34%) and T.V. (31%) were the most important media. Food accounted for 14%, drinks for 10.6% and agricultural inputs for 0.5% of total expenditure on publicity in 1987.

Within promotional expenditures for trade brands, food products accounted for 10.3%, drinks for 5.5% and agricultural inputs 3.2%.

THE PERFORMANCE
OF THE MARKETING SYSTEM

Competitiveness may be considered one of the most significant dimensions of performance, especially after the events in the centrally planned economies.

In Spain, competition policy has been in the past a by-product of the market economy with only marginal attention from public and private institutions. However, the integration in the EC is changing the panorama, amongst others by the penetration of products from other EC countries in the domestic market and by new regulations.

The competition in the agro-food sector has been focused on three main scenarios: low prices, new products and brands, and different qualities. The center of the market bargaining process has moved from farmers to manufacturers, and later on to distributors. The concentration process in food industry, and the appearance of new retailers influence competitive strategies. Although these developments are common to other European countries, in Spain the adaptation period has been shorter, with consequently greater efforts and risks for entrepreneurs.

Bargaining power of retail chains has increased significantly during recent years. Some food factories depend upon hypermarkets for 70% of their product sales. This development in favor of hypermarkets is strengthened by white brands, being used as private brands and in this way increasing the bargaining power versus suppliers.

The attitude of government administration has been to consider the food sector as subject of general economic policy. The first regulation against restrictive practices in commercial activities was introduced in 1963 ('Ley Antimonopolio'). Two different institutions were created: a court (Tribunal de Defensa de la Competencia) and an executive service (Servicio de Defensa de la Competencia). The Spanish Constitution, in article 38, recognizes the free market economy and the necessity to improve competition. Integration in the EC and the new Legislation allow a more effective improvement of competition. Special attention should be given to the coordination of national and Communitarian institutions in order to facilitate 'workable competition.'

Another performance dimension is price index on farm and consumer level. The policy to maintain (average increased 0.66% over 1989) nominal prices for agricultural products has originated a decrease in real terms of prices at farm level.

Food product prices (weighing 33% in the general consumer prices index CPI) increased around 6% in 1990, above the average in the EC and OCDE countries.

The increasing price of land, as part of the production cost of agricultural products, is worth mentioning. In recent years there has been an upward trend, while in the other European countries prices have been stable. High land prices in combination with the increase of labor costs are diminishing Spanish competitiveness in international markets. The negative trend in competitiveness is disadvantageous to Spain. Traditionally, Spain has two kinds of agriculture according to their final orientation. There is a 'continental and northern' agriculture (cereals, meat, milk) oriented toward the domestic market. The so called 'mediterranean agriculture' is more competitive in foreign markets (wine, fruit and vegetables, olive oil).

Food industry has been concentrating on the domestic market because of its low competitiveness. The foreign market was used as secondary outlet, to dispose of surpluses, with some exceptions. Protectionism and lack of export incentives bring about that few enterprises are successful in foreign countries.

MARKETING CHARACTERISTICS OF MAIN CROPS

The analysis of marketing in different agricultural sectors should pay attention to the economic situation of Spain before the process of joining the EC.

There is a group of products, with a relative competitive position because of comparative production advantages (olive oil, wine, fruit, and vegetables). Domestic prices were lower than in other European countries. Because of certain problems with production in the past, cooperatives and other producer organizations are important. Associations of first and second degree, with vertical integration of production and marketing, do control a great percentage of the market.

For other products (cereals, milk) the situation is different. With relatively high production costs and structural marketing problems, the possibility to compete is very low. Only geographical advantages may allow them to survive by means of selling their products at the domestic market.

Poultry and hog sectors are well organized both in production and marketing. Vertical integration accounts for a great part of business (up to 80%), including contract farming, where input and output prices are negotiated for certain periods of time. However, they do have disadvantages in comparison with other European entrepreneurs. Compound feed is more expensive (sometimes up to 20%) because of higher cereal prices and lower use of substitutes. The marketing channels are similar to other European countries.

Cattle is a problematic sector because of comparative disadvantages in production costs. Traditional marketing channels from rural markets to community slaughterhouses have changed, giving more importance to private firms.

With respect to 'cash crops' (sugar, tobacco, oilseeds) the transformation process has stimulated enlargement of enterprises, sometimes resulting in a geographical monopoly.

Marketing channels were controlled in the past either by private or by state companies and open competition was very low. Technical and organizational innovation in the marketing channel is necessary in most sectors. Foreign capital may change the situation in the near future.

One aspect we should consider in this chapter is the performance of the Spanish Administration. The bureaucracy has been suffering under a very controversial transformation. On the one hand there is a 'decentralization activity' as a consequence of the new regional administration process (the 'Autonomias'). Thus, the regions now have some administrative responsibilities, formerly performed by the National Administration. On the other hand there is a 'centrifugal process' towards the EC Administration in Brussels. As a result many entrepreneurs are disoriented, and try to adapt to the new situation, while facing high competition and lack of adequate information.

The Treaty of Spains' integration in the EC has been another point of friction in the Spanish food industry. As a consequence of political and economic pressures, some sectors, such as fruit and vegetables, have been, in their opinion, discriminated in favor of others. After the Spanish integration in the EC, agreements with the USA, some Latin American and Eastern European countries, allowed imports of agricultural products at very competitive prices in the Spanish market, without adequate information for the entrepreneurs.

Spanish farmers should try to orient their production towards the changing needs of consumers in the European market, and to provide adequate raw products to food processors. A good quality-price ratio and integration through contractual arrangements may diminish price instability in the future. Following this, one of the solutions is to encourage vertical contracts, especially for products not subject to the Common Agricultural Policy (CAP). Accordingly, it is estimated that more than one hundred thousand farmers signed contracts under the Farm Contract Law in 1987.

Different forms of ownership in the food sector may facilitate the performance of the integration process: joint ventures, international licensing, franchising, and management contracts.

CONCLUSIONS

Agricultural marketing in Spain is becoming the cornerstone in the evolution process of agrofood sector. Within a transforming economy, markets are being forced towards more dynamic action. However, the market structure is still inadequate, with a predominance of small traditional enterprises.

Competition is becoming one of the most important instruments in the changes, although it may be a source of discrimination. Thus, open competition exists in the markets for final products (food for direct consumption) but not in the input sector (energy, labor, financing facilities, etc.).

Changes may be more radical in the near future, especially in the market structure, because further concentration in the food industry and distribution systems can be expected.

The performance of different sectors has been very uneven. Farming has been strongly dependent on CAP (Common Agricultural Policy) regulations. Food industry is suffering very strong competition. Distribution has been the most dynamic sector, with very liberal regulations for setting up new enterprises, capital market, innovation and other activities.

REFERENCES

Bain, J., 1968. Industrial organization. John Wiley.

Briz, J., 1972. Aplicacion de estudios de mercados al caso de la cebolla en la region valenciana. REAS no. 81, Madrid.

Briz J. and J. Casares, 1985. Mercados agroalimentarios y formas comerciales en la Española Economia y Sociologla Agrarias.

Caldentey, P., J. Briz and others, 1987. Marketing Agrario. Editorial Agricola.

Herrero, A., 1976. Comercializacion asociative de productos agrarios en España. REAS, no. 94.

Jordana, J., 1990. El resistible declinar de la industria alimentaria. Economistas, no. 47. Anuario de la distribution en España, 1990.

Mortes, V., 1990. Las empresas de alimentacion: problematica actual y posibles orientaciones estratégicas. Economistas, no. 47.

Padberg, D. I. and Thorpe, 1974. Channels of grocery distribution: changing stages in evolution. A comparison of USA and UK. Journal of Agricultural Economics, Vol. XXV, no. 1 January.

Scherer, F. M., 1970. Industrial market structure and economic performance. Rand McNally.

Characteristics
of Agricultural Marketing
in Hungary During the Formation
of Market Economy

József Lehota

SUMMARY. Although the central planning system of the Hungarian economy changed after 1968, there was not any considerable decrease of state interference nor was there any significant increase in the role of the market. It was the system of state interference which changed: instead of direct methods the emphasis was primarily put on indirect methods (price subsidies, taxes, loans). The role of the market continued to remain of secondary importance, which was mainly indicated by the low level of market competition. In addition, the economic system fundamentally affected agricultural marketing. The role of the market started to be valued significantly from 1989-1990. In this article I shall describe the initial situation of a market-oriented economy as well as the major trends and the first steps of the new programme.

INTRODUCTION

Although the central planning system of the Hungarian economy changed after 1968, the role of the market continued to remain of

---

József Lehota is Associate Professor, Head of Department of Agricultural Marketing, Gödöllô University of Agricultural Sciences, Gödöllô, Hungary.

[Haworth co-indexing entry note]: "Characteristics of Agricultural Marketing in Hungary During the Formation of Market Economy." Lehota, József. Co-published simultaneously in the *Journal of International Food & Agribusiness Marketing* (The Haworth Press, Inc.) Vol. 5, No. 3/4, 1993, pp. 179-195; and: *Food and Agribusiness Marketing in Europe* (ed: Matthew Meulenberg), The Haworth Press, Inc., 1993, pp. 179-195. Multiple copies of this article/chapter may be purchased from The Haworth Document Delivery Center [1-800-3-HAWORTH; 9:00 a.m. - 5:00 p.m. (EST)].

secondary importance, which was mainly indicated by the low level of market competition. In addition, the economic system fundamentally affected agricultural marketing. The role of the market started to be valued significantly from 1989-1990.

THE CHARACTERISTICS OF FOOD CONSUMPTION

The level of foodstuff consumption in Hungary is relatively favorable from the point of view of dietetics. The energy content of food consumption is 14,100 KJ/person/day (1989). Protein consumption is 106 g/person/day (60 g of it is animal protein), carbohydrate consumption is 400 g/person/day and fat consumption is 150 g/person/day. The division of food consumption according to groups of products is as follows: meat and meat products 78 kg/year, milk and dairy products 189 kg/year, eggs 19.5 kg/year, vegetable and animal fat 37 kg/year, cereals 109 kg/year, sugar 39 kg/year, potatoes 53 kg/year, vegetables and fruits 155 kg/year.

The annual increase in food consumption per capita (in terms of expenditure) varied between 1.5 and 3.3%/year between 1951 and 1975. From 1976 onwards the rate of growth significantly decreased (1976-1980: 0.1%/year, 1980-1985: 1.0%/year). Since 1988 the volume of food consumption has continued to decrease.

The amount spent on provisions accounts for 1/3 of individual expenses, disregarding spending on stimulants (at 1989 prices). Spending on provisions makes up 29.9% of total expenses in households with one wage earner, and 39.8% in households with pensioners.

Retailers play the most important role as sources of acquisition of foodstuffs (1989: 64.4%). Local markets have a share of 8.1% and the share of companies, institutions and other sources is 10.4%. Seventeen and one-tenth percent of food consumption comes from self-production.

The most important factors influencing food consumption in the future will be:

- The 1.2% decrease in the population of Hungary since 1981 due to a decreasing birth rate and an increasing death rate. In 1989 the Hungarian population was 10.6 million.

- The age structure of the population which shows that the proportion of people over 60 is increasing (1960: 13.8%, 1989: 18.7%), and the proportion of people under 29 is decreasing (1960: 50.2%, 1989: 40.4%).
- The growing proportion of urban population (1960: 51.3%, 1989: 61.9%). Thirty-one and nine-tenths percent of the population live in the ten largest cities of Hungary, including 19.4% in Budapest. However, 57.5% of small villages have fewer than 1000 inhabitants (their proportion of the total population is 7.6%).
- The decrease in the average size of family households, from 3.1 persons in 1960 to 2.8 persons in 1989.
- The increase in the proportion of one-person households (1960: 14%, 1985: 20%).
- During the last few decades, with the increase in the employment of women, the proportion of households having two or more wage-earners likewise increased. This process has stopped for the time being.
- The differentiation of the income of the population is relatively low. The poorer 30% of the population earned 17.4% of incomes, the middle class, about 40% of the population, earned 36.1% and the wealthier class (30.0%) 46.5% of incomes (1987).
- The equipment of households is favorable particularly with regard to numbers of refrigerators (104 units/100 households), the numbers of deep freezers are moderate (44 units/100 households), but in the area of microwave ovens and other kitchen equipment numbers were very low in 1990 but have risen sharply in the last two years.

EXPORTING AND IMPORTING AGRICULTURAL FOODSTUFFS AND THE EFFECT ON THE HUNGARIAN ECONOMY

The export of agricultural and food-products has traditionally played an important role in the foreign trade turnover. It constituted 21.6% of the total export between 1976-1980, 22.3% between

1981-1985 and 20.8% between 1986-1990. Up to 1991 the clearance of export accounts was characterized by two systems: one conducted in convertible currencies and the other in roubles. The clearance with the ex-Comecon countries was partly done in dollars, partly in roubles (the proportion of clearance in dollars was: 31.5% between 1976 and 1980, 28.6% between 1981 and 1985 and 26.4% between 1986 and 1990).

The proportion of agricultural exports to total exports was 32.1% in 1990. Unprocessed or semi-processed food products accounted for a relatively high share (primarily in rouble clearance).

The major features of the export of agricultural and food products are as follows:

- the volume and structure of exports are basically determined by domestic production. Since 1976, which marked the start of a decrease in rate of domestic food consumption, the role of exports has continuously increased. This increase is not explained by a development in marketing strategies, it is rather the result of the availability for export of a domestic surplus. This process was reinforced by the increase in our obligations concerning loan repayment to the ex-Comecon countries, primarily the Soviet Union (according to some contracts, e.g., food for energy).
- Hungarian agricultural and food products are exported to some 96-98 countries. During the last few decades two regions have assumed a determinant role: first, the EEC countries. At the beginning of the 1970s markets in EEC countries predominated and from the mid-70s the proportion of exports to ex-Comecon countries was continuously growing. In 1982 nearly 2/3 of agricultural and food product exports were aimed at the Soviet Union.

In 1990 the two regions accounted for nearly identical proportions of Hungarian agricultural exports. Among EEC countries our most important markets are in Germany, Italy, France, the Netherlands and Belgium, and among non-EEC countries, Yugoslavia, Switzerland and Sweden. Among the ex-Comecon countries our main partners are the Soviet Union, Czechoslovakia, Poland and Rumania. The share of non-European countries is low: North and

South America 6%, Africa 0.5%, the Middle East 2%, the Far East 1%. The concentration of exports is relatively strong. The share of CR-4 (Soviet Union, Germany, Italy and Yugoslavia) is 56.9%, the CR-8 indicator (the above countries, plus Austria, Rumania, the USA and France) is 72.3%. Since 1991 exports to the ex-Comecon countries, primarily the Soviet Union, have significantly decreased.

- The export of agricultural and food products includes nearly 600 products. In our most important export markets the choice of goods shows considerable diversity. Exports to the ex-Soviet Union comprise 16% of our products, to Germany 60%, to Italy 36%, and to Yugoslavia 34%. In the two main regions, the choice of products and their level of processing show a significant divergence. The EEC countries import from Hungary primarily highly-processed, and processed products as well as special products (e.g., processed feathers, goose liver, seed grain, honey snails, etc.). In exports to the ex-Comecon countries the proportion of traditional, slightly-processed mass-products dominates. In recent years 80% of the exports to the Soviet Union were composed of five products (1989: wheat, beef-cattle, sugar, butter, 1990: wheat, beef-cattle, pork, chicken).

Between 1976 and 1980 the proportion of imported agricultural products and foodstuffs was 8.3%, between 1981 and 1985 it decreased to 6.5% and between 1986 and 1990 it rose to 7.2%. If industrial equipment necessary for the production of the agricultural and food industries is also included, the proportion of the import will be 15-16%. As a percentage of all imports the rouble clearance import increased from 14.1% between 1976 and 1980 to 18.9% between 1986 and 1990. Twenty-nine and nine-tenths percent of the imports were agricultural products and 70.1% were foodstuffs.

In the Hungarian economy food self-sufficiency has been a characteristic feature for a long time. Therefore, imports can be divided into two major groups: products that cannot be produced in Hungary (tropical fruits, spices, sea fish, fish meal, seed grain, breeding animals), and products that cannot be profitably produced (soya, rice and other protein feed).

Recently, the role of the so-called variety enlargement import products has strengthened. It has appeared mainly in the field of beverages, tobacco products and the confectionery industry. The bulk of the imports of these agricultural products is composed of raw coffee, cocoa (40.3%), fresh vegetables and fruits (23%) and cereals (19.3%). Among food products, vegetable oil products dominate, mainly soya groats (42.8%), beverage and tobacco products (21.1%) and meat, poultry and dairy products (18.8%).

In 1987 the progressive liberalization of the import sector was undertaken. The range of liberalized products in the Hungarian economy sharply increased from 42% in 1989 to 80% in 1991. The proportion of liberalized agricultural products was 36% and that of food industry products was 60% of the total 1990 turnover. As a result of negotiations with the EEC and GATT further liberalization can be expected.

GOVERNMENT POLICY FORMATION

Although Hungary abandoned its traditional system of central planning in 1968, the establishment of a real market economy was slow. State interference in the economy remained considerable. The main forms of interference included the central control of both producer and consumer prices in the food industry sector, the widespread system of subsidies to producers, the processing industry and consumers, strong centralization of incomes through the tax system, and the state system of credit financing.

The comparison of agricultural producers' input and consumer prices in the EEC and Hungary is as follows according to S. Mészáros and M. Spitalszky (1991): in 1989 the prices of certain Hungarian-produced products were similar to those of the EEC (pork 94%, of EEC prices mutton/lamb 79%). In the second group of products (rice, beef, milk, eggs) the Hungarian prices were 51-60% of the EEC prices while in the third group the difference was the largest. Here the prices of Hungarian-produced cereal products were only 37-42% of the EEC prices. The low prices of

Hungarian products are the result of relatively low agricultural input prices.

The foodstuff prices for Hungarian consumers were only 15-52% of the average EEC prices (1987, based on a comparison including 7 products). The biggest divergence appeared in the consumer prices of beef, milk, cheese, the smallest difference was in the consumer price of sugar. Of course, there are several other divergences in the price system of the EEC countries and Hungary, which will not be dealt with here.

In 1988 the net PSE index of Hungarian agriculture was + 18% and in 1989 it was −3%, which was considerably behind the EEC average. The structure of the subsidy systems of Hungarian agriculture (É. Borszéki, 1991) significantly differs from the subsidy system of the EEC. The PSE indices of products in 1980 were between −108 and + 50%, in 1985 between −80 and +57% and in 1989 between −117 and 59%.

Heavily subsidized products include the following: sugar (up to 1988), poultry, pork, eggs and mutton. In plant production the PSE index of cereals is negative. On the basis of the data of 1986 and 1988 the subsidy level of pork, poultry, eggs was higher than in the EEC.

The third area of state interference is the tax system. In Hungary the rate of depletion of company incomes is high by international standards. Various taxes make up 62.4% of the GDP in the average of the national economy (1989) (É. Borszéki, 1991). The rate of taxation on the food industry is higher and that on agriculture is lower than the average.

Parallel with the formation of a market economy, the system of state interference has also to be changed. The elaboration of the system of agricultural regulation is currently being undertaken and it is expected to be introduced in 1992.

Initially, this agricultural regulation will apply to cattle and pigs for slaughter, cereals, and milk. The most important areas of agricultural market regulation will be the following:

- price and competition regulations,
- quantitative regulation of agricultural production,
- export and import regulation,

• the system of intervention,
• the system of agricultural subsidization.

THE FORMATION OF A FREE-MARKET STRUCTURE

Agricultural production is strongly polarized in Hungary. At one end of the spectrum there are 131 over-centralized state farms with an average territory of 7,315 ha and 1,242 agricultural cooperatives with an average territory of 4,208 ha.

At the other end of the spectrum there are approximately 1.5 million strongly decentralized small farms with an average territory of 0.52 ha. Fundamentally, large-scale farms can be well-mechanized, dealing as they do with cereal, sunflower, sugar beet and vegetable production. Small-scale farms deal with labor intensive vegetable and fruit production, which is also difficult to mechanize. Among branches specializing in animal husbandry, small-scale farming is predominant in pig farming and in raising small animals (e.g., rabbits), it has nearly the same role in poultry husbandry and a somewhat smaller role in cattle and sheep husbandry. Crops are predominantly grown on large farms. The average size of wheat fields on state farms is 1,329 ha/farm, and 873 ha/farm on co-operative farms, in corn production the average size is 1,154 ha and 641 ha, in sugar beet production it is 365 ha and 230 ha, in sunflower production it is 419 and 335 ha (1988).

Animal husbandry in agricultural large-scale farms is just as concentrated: the stock of cattle is 3,046 and 1,054 animals/farm, that of pigs is 18,279 and 4,536 animals/farm, and that of sheep is 3,932 and 2,618 animals/farm.

On average small-scale farms had 3.3 cattle, 5.5 pigs and 14.2 sheep (1986). The size of plant producing branches was equally low. The proportion of potato producing small-scale farms larger than 0.5 ha, was 0.3%; the proportion of farms raising more than 10 cattle was 1.4% while the proportion of farms raising more than 100 pigs was 1.7%, and the proportion of farms raising more than 100 sheep was 27.4%. Presently, medium-sized farms have completely disappeared from Hungarian agriculture. The privatization of state farms, the re-structuring of agricultural co-operative farms and the compensation of previous land owners have started. Due to this process the over-concentration of large-scale farms is expected to decrease, whereas the size of small-scale farms is expected to increase.

The production of the Hungarian food industry is conducted primarily in two forms: in the traditional food industry companies as well as in the food processing plants of state farms and co-operative farms. The number of food processing plants of large-scale farms was 3,408 (1985).

The market structure of state-run food-industry companies can be characterized by the following features: firstly, by a centralized company structure, secondly, by the fact that for a long time they operated in a so called "limited special line," thirdly, by the fact that they are organized according to sectors (mainly in the dairy, cereal and meat industries).

From the 1950s the organizational system of the state-run food industry was generally highly centralized, and composed of two forms, namely, national companies and trusts. The state-run companies in the food industry were grouped into five trusts and ten national companies (1964). These forms survived until the beginning of the 1980s. Between 1980 and 1982, 7 out of 15 organizations were transformed into individual companies. In 1991 there was only one large national company. In this horizontally organized monopoly structure no significant competition functioned either in the collection of agricultural products or in their sale. In the domestic market competition existed primarily between the state-run companies of the food industry on one hand and agricultural large-scale farms on the other. During the previous decades the role and market share of the private food industry was low. In 1987 the number of private food producers was 4,130.

The market concentration of state-run food-industry companies was high. According to 1989 figures the degree of concentration was as follows:

- trust and national companies existed in the milk, grain and vegetable-oil industries,
- the rate of the CR-4 index was above 75% in the beer, tobacco and confectionery industries,
- the CR-4 index was between 51 and 70% in the poultry and sugar industries,

- the CR-4 index was between 30 and 50% in the conserve industry (canning and cooling industries).

The wholesale and retail of agricultural products and provisions were characterized by the following:

- the relatively sharp division of profiles (wholesale trade-retail trade, domestic trade-foreign trade),
- the dominance of operations according to the principle of area division,
- the lack of variety in marketing and in channels of distribution.

The wholesale marketing of agricultural products was basically carried out by food processing plants of state-run and agricultural large-scale companies.

In food processing the dominant state-run food-industry companies collected and processed agricultural products or passed them on to other state-run food-processing plants basically according to the principle of a division structure.

In the wholesale marketing of potatoes, vegetables and fruits, excluding the canning and cooling industries, specialized wholesale companies (county companies of ZÖLDÉRT) as well as ÁFÉSZ (271 General Consumers' and Marketing Co-operatives) played significant roles. There were also special wholesale companies, e.g., in the collection of sowing seed, tobacco, wool and sheep.

In recent years there has been a significant increase in the number of private enterprises specializing in the wholesale marketing of agricultural products. Their market share is continuously growing, but it is still relatively low.

In the purchase of products of small-scale farms (vegetables, fruits, potatoes, pigs, rabbits, etc.) in addition to ÁFÉSZ Co-ops and agricultural co-operative farms played an important role as well.

The wholesale food trade is partly conducted by the food-industry companies (the baking, milk, meat, poultry, beer and distilling, wine and tobacco industries). The marketing of other products is conducted by companies having functions of wholesale trade and which were organized according to the principle of area division (grain, canning, vegetable oil, sugar and confectionery industries). State-run food-industry companies were also involved in marketing the products of other food-industry companies. There were separate wholesale companies for certain products or areas, e.g., in the case of the meat and poultry industries, and the products of the tobacco industry. In recent years private enterprises have been established to deal with the wholesale marketing of foodstuffs, but their market share is still low.

State-run companies, co-operatives, and private retail firms also take part in the retail marketing of foodstuffs. The state-run retail trade is also formed according to the principle of area division and has operated for a long time, e.g., KÖZÉRT companies in Budapest, country retail companies in country towns. In villages and smaller towns the role of ÁFÉSZ Co-ops was dominant in the retail trade of foodstuffs. Although this principle of areas division has recently become less important, its influence can still be felt. The number of national commercial chains is low in the retail food trade (SKÁLA-COOP, Csemege-Meinl as well as the Budapest KÖZÉRT Company operating throughout the city). The number of food retail shops was 22,537 in 1989, with 50.5% of them being under private ownership. The proportion of private retail outlets in the turnover of retail foodstuff is 7.2%. The number of private retail shops has considerably increased since 1989, but predominantly small-sized shops have been established with a limited variety of goods.

Privatization has started in the case of both the state-run food industry and the state-run wholesale and retail food trade.

In the privatization process there are no limitations on the amount of domestic or foreign capital. Between 1980 and 1988 the major forms of cooperation between domestic and foreign food-industry companies were in the purchase of licenses. At the end of 1990 there were 130 joint ventures employing foreign capital in the state-run food industry, 78 in agriculture and 1,661 in internal trade. In 1991 this process accelerated, in the first six months 69 additional ones were established in the food industry, 42 in agriculture and 1,123 in internal trade.

In the food industry the participation of foreign capital has been considerable in the sugar and confectionery industries, but joint ventures are also being set up in the vegetable oil, beer and tobacco industries. The establishment of joint ventures operating with foreign capital has likewise started in the wholesale and retail food trade.

MARKET ORGANIZATION

In the state-run food industry and the wholesale and retail food trade, the horizontally organized monopoly organizations, operating according to the principles of area division, were dominant for a long time.

The wholesale and retail food-processing activities of agricultural large-scale plants, and the role of the co-operative and private retail trade have been continuously increasing. The principles of area division have declined in importance but they have not disappeared.

To counterbalance this monopolistic situation the state controlled both the producers' prices of agricultural products and those of the food-industry's products as well as the profit margins of the wholesale and retail food trade. Despite the gradual liberalization, the marketing of nearly 2/3 of agricultural products and foodstuffs was conducted until 1989 at centrally determined and controlled fixed prices. There were fixed profit margins both in the wholesale and retail trade. Administrative regulations also played an important role in the wholesale and retail food trade. The liberalization of prices and the cessation of market obligations had become general by 1990.

Due to these characteristics there was an absence in Hungary of a considerable number of market organizations which are typical of developed market economies. Because of the special features of agricultural co-operative farms, the traditionally known marketing co-operatives did not exist. The ÁFÉSZ Co-ops and special groups of agricultural co-operative farms were similar to marketing co-operatives.

These groups absorbed the small-scale producers. In 1987 2,681 special groups were in operation. Each group conducted the sale of products of 79 part-time small-scale producers.

The groups primarily specialized in the sale of certain products, e.g., rabbit meat, grape, fruit, vegetable, and other agricultural products. These special groups purchased brood animals, feed, artificial fertilizers, herbicides, pesticides, machinery and other materials as well.

Currently, the horizontal marketing organizations of agricultural producers (organizations similar to marketing boards) do not exist. Horizontal organizations which are now being formed are mainly producers' associations.

Although at present the role of market organizations is not substantial (e.g., auctions, wholesale markets), they are important from the point of view of the fixing of the prices of agricultural producers. Sales at auctions are extremely low (e.g., the sale of racehorses and brood animals). Wholesale markets exist in two areas: in the sale of fruit and vegetables and in the sale of flowers, both located in Budapest.

In Hungary wholesale markets are the typical marketing channels of small-scale firms, especially in vegetable and fruit marketing. Although the price forming and information roles of wholesale markets currently exist, functions of standardizing, quality control and promotion have practically not yet been realized.

The Budapest Commodity Exchange Ltd. was established in 1989, which resulted in the liberalization of grain prices. During its first year of operation the BCE dealt with cereals, and from 1992 has expanded its scope to include the meat sector as well. Presently, wheat, maize, rye, barley, sunflower and pigs are products quoted on the Commodity Exchange.

Currently, the functions of Agricultural regulation and intervention are not yet operational, but the Commodity Exchange still plays an important role in price formation and price information. In 1990 the turnover of cereals on the Commodity Exchange was equal to 1-1.5% of the total turnover of cereals.

THE BASIC FEATURES OF AGRICULTURAL MARKETING

Due to the absence of a market economy and strong state intervention we can only talk about the most elementary features of agricultural marketing. The establishment of a market economy and the reduction of state intervention has only a very short history.

Previously, the agricultural and food industries were totally production-oriented.

Quality only became more important in the 80s when Hungarian exports began to have problems with sales abroad.

The low level of competition was seen particularly in the lack of product differentiation and innovation. In agriculture the application of new species and technological innovation was accelerated after the 60s. Development was considerable in grain, sunflower and sugar beet production, on one hand, and in poultry and pig breeding, on the other hand. It was less important in vegetable and fruit production as well as in the sheep breeding branch.

In the food industry from an international perspective, there was little product development. (Mrs. G. Várhelyi, A. Tükrössy, 1989; P. Gergely, 1988). In the mid-80s the average age of products in the state-run food industry was 22 years, with the proportion of products introduced before 1950 being 26%. The proportion of new products (1-6 years) was 15%. The proportion of new products was even more unfavorable in the return from sales (0.3-0.5% on the basis of the average return of 1982-87). The situation was relatively more favorable in the confectionery, distilling and tobacco industries. The proportion of R+D expenses as a percentage of profit was only 4.6% (26% of the industrial average).

The use of brand names for Hungarian agricultural products was minimal, in the case of foodstuffs certain brands dating back to the pre-war period survived (Hertz, Pick, wines, drinks). Few new products were given brand names, except in the case of licensed products and in branches where import competition was strongest. Here industrial brands were dominant, commercial brand names were less significant.

Quality regulation and control of agricultural products and foodstuffs is basically centered on standards. The foodstuff law includes the general regulations of food production, storing and marketing. Quality control, also centered on standards, has a vertical aspect. Elements of horizontal standardization cover certain components of human and animal hygienic, chemical-microbiological contamination.

Standards of agricultural products cover sensory qualities (e.g., size, weight, taste, ripeness, homogeneity) as well as qualities of ingredients (e.g., milk fat, milk protein, dry material, sugar and oil content) and microbiological qualities.

The quality control of agricultural products is supported by the registration system of brood animals and by animal and plant hygiene control. Standards relating to agricultural products are primarily producer oriented. Due to the prevailing fixed price system a tight relationship has not been formed between price and quality.

The quality control of provisions is also based on standards. The quality control of sensory features, macro components, is regulated, but there is a lack of control of micro components (amino-acids, sebasic acids, vitamins, etc.) as well as of that of microbiological components. The present system of quality control is basically final product oriented.

The most important current changes in the quality control system of agricultural products and provisions are as follows: firstly, the strengthening of horizontal regulations by the elaboration of Hungarian food legislation. Secondly, the adaptation of our quality regulation system to that of more developed countries, primarily to that of the EEC. Thirdly, the strengthening of consumer oriented attitudes towards quality control. Fourthly, the improvement of the technical conditions of quality control.

The price system of agricultural products has been based on fixed or controlled prices for a long time. In 1986 there were fixed prices for wheat, milk, cattle for slaughter, red pepper spice; there were controlled prices for maize, sunflower, apples, pigs for slaughter, potatoes, vegetables, fruits and poultry for slaughter. The fluctuation of fixed and controlled prices depended on the decisions of state authorities since market influences had very little effect. A smaller differentiation of prices occurred depending on quality and demand. The food-industry producers' prices and consumers' prices were similar to agricultural producers' prices. Price liberalization occurred at the end of the 80s and at the moment the prices of agricultural products are freed. Within the framework of agricultural market regulations (expected sometime in 1992), a new price system will be introduced regulating the most important agricultural products.

Between 1980 and 1989 the farm price index was 160.3%, the

non-farm price index was 165.4% and the consumer's food price index was 206%.

Another important feature of agricultural marketing was the low number of marketing channels. This was relevant for both domestic and export marketing. In 1986 there was only one marketing channel (P. Tomcsányi, 1988) in the marketing of wheat, sugar beets, sunflower, cattle for slaughter, milk and seasoning paprika. In the marketing of other products the number of channels was low. The administrative limitations on the choice of marketing channels have been stopped, but a real multi-channelled marketing system has not yet been established.

Competition, among home-produced products, import competition, and promotion systems are relatively under-developed because of the administrative limitations of marketing channels and the price system. In Hungary the proportion of the amount spent on display advertising is 0.13% of GDP (1990), which is only 20% of the West-European average.

The amount of display advertising expenditure was US $2/person, which was only 1.3% of the West-European average. There was an even less favorable situation in the case of agricultural products and foodstuffs. In advertising the estimated share of the media is as follows: the proportion of newspapers is 25-30%, that of television is 20-25%, that of brochures, leaflets and company publications is 20-25%, that of street posters is 5-10% and that of radio is 3-5%. In the case of foodstuffs there was stronger promotional activity in industries where there was stronger competition. Due to the lack of co-operative agricultural marketing bodies advertising activity of this kind is completely non-existent. During the past couple of years the advertising market has become more lively due to the widespread emergence of joint ventures and foreign advertising agencies.

CONCLUSION

The first phase in the formation of a market economy has caused radical changes, which can be seen on one hand in the freeing of prices, farm and consumer prices, as well as the decreasing of food subsidy, and on the other hand in the strengthening of domestic and import competition. This process has been accompanied by significant changes in agricultural and food industry markets: the decrease in domestic food consumption, the loss of some traditional export markets, and the opening of some new ones. Besides market factors, significant changes have been taking place in the privatization of state farms and food industry companies: in the restructuring of old-fashioned co-operatives and in the formation of new farm structures. These changes have led to a quick and significant reduction in the bargaining power of agricultural producers. To change this situation, two factors are essential: first, the strengthening of the market orientation of agricultural producers, in part through the formation of marketing organizations, for example marketing boards and co-operative marketing organizations; second, by the state's role in the formation of a market support system and by the operation of subsidy systems.

REFERENCES

Árgyelán Gné (1984). Az állami élelmiszeripar és belkereskedelem szervczeti rendszerének kapcsolata. Gazdálkodás, 3. sz. 33-40. p.

Biró Oné (1986). Az élelmiszertemelés és forgalmazás irányí-tási és szervezési rendszere. Gazdálkodás, 8. sz. 56-62. p.

Borszéki É. (1991). A mezôgazdasái adórendszer az Európai Kö-zösséghez közeledés jegyében. Gazdálkodás, 6. sz. 26-34 p.

Borszéki É. (1991). A mezôgazdasái termelés és termelôk pénz-ügyi támogatása az Európai Közösségben I. Gazdálkodás, 6. sz. 13-25. p.

Gergely P. (1988). Innovácuó az élelmiszeriparban. Élelmezéssi ipar, 9. sz. 335-338. p.

Mészáros S., Spitalszky M. (1991). Az Európai Közösség és Magyar-ország árrendszerének összehasonlítása. Gazdálkodás, 6. sz. 1-12. p.

Szabó J. (1991). Agrárkülkereskedelmünk prioritásai. Élelmezési ipar, 7. sz. 243-247. p.

Szôke M. (1986). A nemzetközi termelési együtmôködés alaku-lása és ökonómiai értékelése az élelmiszer-iparban Élelmezési ipar, 11. sz. 402-405. p.

Tomcsányi P. (1988). Az élelmiszer-gazdaság marketing alapjai. Mezôgazdasági Kiadó, Budapest.

Várhelyi Gné, Tükrössy A. (1989). A marketing szemléletmód érvé-nyesítésének lehetôségei az élelmiszeripari vállalatoknál. Élelmezési ipar, 1. sz. 21-24. p.

Zacher L. (1989). Nagy remények - kis eredmények az élelmi-szeriparban. Gazdálkodás, 5. sz. 29-41. p.

Index

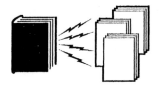

Haworth
DOCUMENT DELIVERY
SERVICE
and Local Photocopying Royalty Payment Form

This new service provides (a) a single-article order form for any article from a Haworth journal and (b) a convenient royalty payment form for local photocopying (not applicable to photocopies intended for resale).

- *Time Saving:* No running around from library to library to find a specific article.
- *Cost Effective:* All costs are kept down to a minimum.
- *Fast Delivery:* Choose from several options, including same-day FAX.
- *No Copyright Hassles:* You will be supplied by the original publisher.
- *Easy Payment:* Choose from several easy payment methods.

Open Accounts Welcome for . . .
- Library Interlibrary Loan Departments
- Library Network/Consortia Wishing to Provide Single-Article Services
- Indexing/Abstracting Services with Single Article Provision Services
- Document Provision Brokers and Freelance Information Service Providers

---

MAIL or *FAX* THIS ENTIRE ORDER FORM TO:

Attn: **Marianne Arnold**
Haworth Document Delivery Service
The Haworth Press, Inc.
10 Alice Street
Binghamton, NY 13904-1580

or **FAX:** (607) 722-1424
or **CALL:** 1-800-3-HAWORTH
(1-800-342-9678; 9am-5pm EST)

PLEASE SEND ME PHOTOCOPIES OF THE FOLLOWING SINGLE ARTICLES:

1) Journal Title: _____

   Vol/Issue/Year: _____ Starting & Ending Pages: _____

   Article Title: _____

---

2) Journal Title: _____

   Vol/Issue/Year: _____ Starting & Ending Pages: _____

   Article Title: _____

---

3) Journal Title: _____

   Vol/Issue/Year: _____ Starting & Ending Pages: _____

   Article Title: _____

---

4) Journal Title: _____

   Vol/Issue/Year: _____ Starting & Ending Pages: _____

   Article Title: _____

(See other side for Costs and Payment Information)

COSTS: Please figure your cost to order quality copies of an article.

1. Set-up charge per article: $8.00

                  ($8.00 × number of separate articles) _____

2. Photocopying charge for each article:

                  1-10 pages: $1.00 _____

                  11-19 pages: $3.00 _____

                  20-29 pages: $5.00 _____

                  30+ pages: $2.00/10 pages _____

3. Flexicover (optional): $2.00/article _____

4. Postage & Handling: US: $1.00 for the first article/

                  $.50 each additional article _____

                  Federal Express: $25.00 _____

                  Outside US: $2.00 for first article/

                  $.50 each additional article _____

5. Same-day FAX service: $.35 per page _____

6. Local Photocopying Royalty Payment: should you wish to copy the article yourself, Not intended for photocopies made for resale, $1.50 per article per copy (i.e. 10 articles × $1.50 each = $15.00) _____

                            GRAND TOTAL: _____

METHOD OF PAYMENT: (please check one)

❏ Check enclosed   ❏ Please ship and bill. PO # _____

                  (sorry we can ship and bill to bookstores only! All others must pre-pay)

❏ Charge to my credit card: ❏ Visa; ❏ MasterCard; ❏ American Express;

Account Number: _____ Expiration date: _____

Signature: X_____ Name: _____

Institution: _____ Address: _____

City: _____ State: _____ Zip: _____

Phone Number: _____ FAX Number: _____

MAIL or *FAX* THIS ENTIRE ORDER FORM TO:

Attn: **Marianne Arnold**
Haworth Document Delivery Service
The Haworth Press, Inc.
10 Alice Street
Binghamton, NY 13904-1580

or **FAX:** (607) 722-1424
or **CALL:** 1-800-3-HAWORTH
(1-800-342-9678; 9am-5pm EST)